PROTEIN FOLDING

PROTEIN FOLDING

A discussion organized and edited by

C. M. DOBSON AND A. R. FERSHT

 CAMBRIDGE
UNIVERSITY PRESS

Published by the Press Syndicate of the University of Cambridge
The Pitt Building, Trumpington Street, Cambridge CB2 1RP
40 West 20th Street, New York, NY 10011–4211, USA
10 Stamford Road, Oakleigh, Melbourne 3166, Australia

First published in *Philosophical Transactions of the Royal Society of London*,
series B, volume 348 (no. 1323), pages 1–119.
This paperback edition published by Cambridge University Press 1996

Printed in Great Britain at the University Press, Cambridge

A catalogue record for this book is available from the British Library

Library of Congress cataloguing in publication data available

ISBN 0 521 57636 9 paperback

Protein folding

A DISCUSSION ORGANIZED AND EDITED BY C. M. DOBSON AND A. R. FERSHT

(Discussion held 18 and 19 October 1994)

CONTENTS

PATHWAYS, KINETICS AND INTERMEDIATES

T. CREIGHTON
Disulphide-coupled protein folding pathways — 5

A. R. FERSHT
Mapping the structures of transition states and intermediates in folding: delineation of pathways at high resolution — 11

SHEENA E. RADFORD AND CHRISTOPHER M. DOBSON
Insights into protein folding using physical techniques: studies of lysozyme and α-lactalbumin — 17

YI WANG, ANDREI T. ALEXANDRESCU AND DAVID SHORTLE
Initial studies of the equilibrium folding pathway of staphylococcal nuclease — 27

O. B. PTITSYN, V. E. BYCHKOVA AND V. N. UVERSKY
Kinetic and equilibrium folding intermediates — 35

ZHENG-YU PENG, LAWREN C. WU, BRENDA A. SCHULMAN AND PETER S. KIM
Does the molten globule have a native-like tertiary fold? — 43

PREDICTION, ANALYSIS AND DESIGN

W. F. VAN GUNSTEREN, P. H. HÜNENBERGER, H. KOVACS, A. E. MARK AND C. A. SCHIFFER
Investigation of protein unfolding and stability by computer simulation — 49

HUE SUN CHAN, SARINA BROMBERG AND KEN A. DILL
Models of cooperativity in protein folding — 61

JANET M. THORNTON, DAVID T. JONES, MALCOLM W. MACARTHUR, CHRISTINE M. ORENGO AND MARK B. SWINDELLS
Protein folds: towards understanding folding from inspection of native structures — 71

STEPHEN BETZ, ROBERT FAIRMAN, KARYN O'NEIL, JAMES LEAR AND WILLIAM DEGRADO
Design of two-stranded and three-stranded coiled-coil peptides — 81

CATALYSTS AND CHAPERONES

KOSTAS TOKATLIDIS, BERTRAND FRIGUET, DOMINIQUE DEVILLE-BONNE, FRANÇOISE BALEUX, ALEXEY N. FEDOROV, AMIEL NAVON, LISA DJAVADI-OHANIANCE AND MICHEL E. GOLDBERG
Nascent chains: folding and chaperone interaction during elongation on ribosomes — 89

RAINER JAENICKE
Folding and association versus misfolding and aggregation of proteins — 97

F. ULRICH HARTL
Principles of chaperone-mediated protein folding — 107

PAUL B. SIGLER AND ARTHUR L. HORWICH
Unliganded GroEL at 2.8 Å: structure and functional implications — 113

PREFACE

The ability of proteins to fold rapidly and efficiently into their intricate and highly specific structures following their synthesis on ribosomes is an essential part of the conversion of genetic information into cellular activity. But in contrast to our understanding of the transcription and translation events, little is understood in detail about how this occurs. In the cell folding is now recognized to be a highly controlled process, with a cascade of proteins involved in ensuring that it occurs in the right place at the right time, and that the newly formed polypeptide chains do not fall victim to unproductive and irreversible events. Many proteins, however, are able to fold *in vitro* in the absence of all of these factors, indicating that the information necessary for folding is encoded in the amino acid sequence. This also allows folding to be investigated *in vitro*, and enables physical and chemical methods to be used to probe the structural transitions involved. The role of individual cellular factors can then be explored by examining their influence on the *in vitro* process, and by comparison with studies of the *in vivo* events themselves.

The meeting on which this volume of papers is based focused strictly on the molecular basis of the folding processes, and brought together a wide range of experimental and theoretical scientists who are leaders in this field. The first group of papers is concerned with the experimental elucidation of the pathways of protein folding. In the opening paper, Creighton discusses the value of studying disulphide-coupled folding pathways, illustrating this particularly with studies of BPTI, a protein with one of the best characterized folding pathways. In the next two papers, Fersht discusses experimental strategies for mapping the structures of transition states and intermediates in folding using protein engineering methods, and Radford and Dobson describe the use of physical techniques, particularly nuclear magnetic resonance spectroscopy, to characterize the structures of species formed during the folding of proteins. The use of spectroscopic methods is discussed further in the paper by Shortle and colleagues who describe approaches to defining folding pathways by studying time averaged structures of denatured proteins under equilibrium conditions. The description of folding intermediates in terms of their thermodynamic as well as structural properties is one of the subjects of the paper by Ptitsyn *et al.*, who discusses the concept of the 'molten globule' state which has had a marked influence on the development of ideas as to how folding takes place. This topic is extended in the paper by Kim and colleagues, in which studies of the structure of one of the classic molten globules, that of α-lactalbumin, by 'protein dissection methods' are described.

The next group of papers is concerned with theoretical approaches to the folding problem, including those which link an understanding of protein folding with the prediction of three-dimensional structures from sequence information and the design of new proteins. Van Gunsteren *et al.* describe the methods of molecular dynamics to probe the unfolding of a variety of proteins in computer simulations, and compare the results of these with experimental findings. Dill and coworkers use lattice models to examine the manner in which the tertiary structure of a protein is encoded in its sequence, and to explore the nature and significance of cooperativity in protein folding. Thornton *et al.* provide a rather different perspective on the folding problem by asking how inspection of the database of protein structures can provide clues as to the interactions that stabilize the final folded states of proteins and how this might provide insight into the folding pathways by which these are attained. Finally in this section, De Grado and colleagues describe synthetic approaches to understanding the structural basis for the interaction of helices in coiled coils, and discuss *de novo* design of proteins based on the rules and concepts believed to be important for protein folding and stability.

The final group of papers address the issue of how proteins fold *in vivo*, and how this relates to the situation found during *in vitro* folding. Goldberg and colleagues address the complex issue of the structural properties of nascent polypeptide chains during synthesis on ribosomes, using a strategy based on the use of conformation dependent monoclonal antibodies. The thorny question of misfolding and aggregation, which can compete with proper folding and association of proteins is addressed by Jaenicke, who also touches on the role of molecular chaperones in preventing and controlling such interactions. This aspect of folding is pursued in detail by Hartl, who describes the hierarchical action of chaperone complexes by which newly formed polypeptide chains are passed from one class of helper proteins to another. The last paper in the volume, by Sigler and Horowitz, describes the crystal structure of one of the most studied of molecular chaperones, GroEL from *E. coli*, and discusses the extent to which this has allowed insights to be gained into its mechanism of action. This paper emphasizes particularly clearly that the description of folding at a molecular level, the theme of the meeting, is no longer the preserve of *in vitro* folders alone.

April 1995

C.M. Dobson
A.R. Fersht

Disulphide-coupled protein folding pathways

T. CREIGHTON

European Molecular Biology Laboratory, D69012 Heidelberg, Germany

SUMMARY

Protein folding pathways that involve disulphide bond formation can be determined in great detail. Those of bovine pancreatic trypsin inhibitor, α-lactalbumin and ribonucleases A and T_1 are compared and contrasted. In each species, whatever conformation favours one disulphide bond over another is stabilized to the same extent by the presence of that disulphide bond in the disulphide intermediates. The pathways differ markedly in the nature of that conformation: in bovine pancreatic trypsin inhibitor a crucial intermediate is partly folded, in α-lactalbumin the intermediates tend to adopt to varying extents the molten globule conformation, while in the ribonucleases the early disulphide intermediates are largely unfolded, and none predominate. In each case, however, the slowest step is formation of a disulphide bond that will be buried in a stable folded conformation; the most rapid step is formation of an accessible disulphide bond on the surface of a folded conformation. Quasi-native species with the native conformation, but incomplete disulphide bonds, can either increase or decrease the rate of further disulphide formation.

1. INTRODUCTION

Determining the mechanism by which a protein folds remains a very difficult and complex task. Small model proteins can fold on a timescale as short as 10^{-2} s, which is very much faster than would be possible by random searching of all conformations, but much slower than the time taken for unfolded polypeptides to interconvert conformations (approximately 10^{-9} s). Therefore, each molecule of an unfolded protein could be sampling up to 10^7 conformations during even very rapid folding. The situation is more complicated for experimental studies, which deal with populations of between 10^{14}–10^{17} protein molecules. Each of these unfolded molecules is likely to have a unique conformation at the start of folding, when the unfolded protein is transferred to refolding conditions, so it is possible that up to 10^{24} conformations could be sampled during folding of all these molecules. Fortunately, the kinetics of folding are relatively simple (Creighton 1988, 1990), suggesting that many molecules adopt the same or similar conformations, following the same pathway, and that relatively few conformations are sampled during much of the timescale of folding. Nevertheless, the partly folded intermediates that are believed to produce rapid folding are difficult to detect, and even more difficult to characterize. In particular, the kinetic roles of the intermediates that are detected are uncertain in most cases, and those intermediates that are kinetically most important are probably not detectable because of their instability (Creighton 1994).

2. DISULPHIDE BONDS AS PROBES OF PROTEIN FOLDING

Many of the difficulties associated with determining protein folding pathways can be overcome if folding is coupled to disulphide bond formation (Creighton 1978, 1986). This is the only type of protein-stabilizing interaction that is susceptible to specific experimental control, due to its oxidation/reduction nature. Folding is coupled to disulphide formation if the folded conformation requires the correct disulphide bonds. The reduced protein is unfolded, and folding accompanies disulphide formation.

To study disulphide-coupled folding of a reduced protein, a disulphide reagent RSSR is added, with or without its reduced form RSH. The spontaneous thiol-disulphide exchange reaction between reagent and protein is monitored as a function of time by quenching the reaction, then separating and identifying the trapped protein species with different disulphide bonds (both those between cysteine residues of the protein and mixed disulphides with the reagent). The kinetics of the folding reaction are measured by varying the concentrations of both RSSR and RSH and by following both folding and unfolding. A plausible kinetic scheme must be able to account quantitatively for the rates of appearance and disappearance of all the protein species, in both directions and with all concentrations of RSSR and RSH. The rate constant that is most pertinent to protein folding is that of the intramolecular step in forming each disulphide bond, in which the mixed disulphide of the reagent with one cysteine residue of the protein is displaced by a second cysteine residue.

It is not sufficient to concentrate on just the most prominent intermediates: an intermediate can be

Phil. Trans. R. Soc. Lond. B (1995) **348**, 5–10

Printed in Great Britain

5

crucial for a kinetic step, yet not accumulate to a significant quantity. Fortunately, the existence of such crucial but ephemeral intermediates can be detected by examining the effects of removing the various cysteine thiol groups (Creighton 1977 a). It is necessary, therefore, to confirm the pathway by studying both the isolated intermediates and the effects of deleting the various cysteine thiol groups of the protein. Finally, the conformations of the trapped intermediates should explain why certain disulphide bonds are made and not others.

3. KNOWN DISULPHIDE FOLDING PATHWAYS

(a) Bovine pancreatic trypsin inhibitor (BPTI)

The most extensively characterized disulphide folding transition is that of BPTI (Creighton 1978, 1990, 1992c), which has been conserved during evolution (Hollecker & Creighton 1983) and is summarized in figure 1a. The kinetic pathway was elucidated by normal kinetic analysis of both unfolding and refolding (Creighton & Goldenberg 1984) and confirmed by removing the various cysteine thiol groups (Creighton 1977a, 1978; Kosen et al. 1992; Darby & Creighton 1993; Darby et al. 1993, 1995). The pathway and the rate constants for all the steps account quantitatively for the rates of appearance and disappearance of all the disulphide species during both unfolding and refolding over a range of redox conditions. The conformational basis of the pathway is now largely understood, because the three-dimensional structures of all the most important intermediates have been characterized in detail (Kosen et al. 1983; States et al. 1987; Darby et al. 1991, 1992; van Mierlo et al. 1991a, b, 1992, 1993, 1994; Hurle et al. 1992; Kemmink & Creighton 1993). These results are incompatible with revisions of the pathway proposed by Weissman & Kim (1991); their results were not inconsistent with any of the important aspects of the original pathway (Goldenberg & Creighton 1984), but largely confirmed it (e.g. Weissman & Kim 1992) and do not require its revision or reinterpretation in any way (Creighton 1992a, b).

Certain intermediates containing solely native disulphide bonds accumulate primarily because they tend to adopt the native conformation; they are indicated by the subscript N. The folded conformation of native BPTI is so stable that it is still populated when any one of the native disulphide bonds is missing. It is remarkable that in these structures essentially all of the residues adopt their native conformations, indicating that the native conformation is a single cooperative structure and that it is only stabilized, not specified, by the disulphide bonds. The fully folded native conformation is still detectable with only the 5–55 disulphide bond (van Mierlo et al. 1991b), which stabilizes the native conformation most (Creighton & Goldenberg 1984). The occurrence of these quasi-native states is decreased by studying disulphide formation at relatively high pH, where it is more closely coupled to folding because the thiol groups tend to ionize, decreasing their tendency to be buried in folded structures.

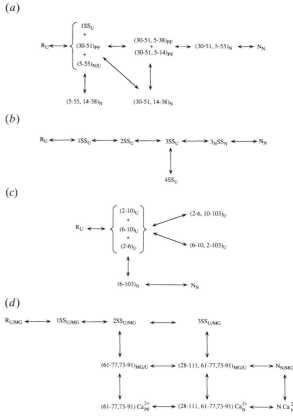

Figure 1. The disulphide folding pathways of: (a) BPTI; (b) RNase A; (c) RNase T_1 and; (d) α-LA. In each case, R is the fully reduced protein. Intermediates with specific disulphide bonds are indicated by the residue numbers of the cysteine residues paired. Mixtures of many isomers with different disulphide bonds (which are usually interconverting by intramolecular disulphide rearrangements when thiol groups are also present) are depicted by the number present, i.e. 1SS, 2SS, 3SS and 4SS. Arrows to and from such species represent the creation, destruction or rearrangement of many different disulphide bonds. The designation 3_NSS indicates that all the disulphide bonds are native-like. N is the protein with all the native disulphide bonds: (30–51,14–38,5–55) in BPTI; (26–84,40–95,58–110,65–72) in RNase A; (2–10,6–103) in RNase T_1 and; (6–120,28–111,61–77,73–91) in α-LA. The arrows indicate the most important steps, in kinetic terms, and can involve either disulphide creation, destruction or rearrangement. The predominant conformations of the various species are indicated by the subscripts U (unfolded), MG (molten globule), PF (partly folded), and N (native-like). Where two or more conformations are present in equilibrium, they are separated by /. The brackets in (a) and (c) indicate that the one-disulphide intermediates included are in equilibrium; the ' + ' between two or more species indicates that they have the same kinetic roles. Taken in part from Creighton et al. (1995); used with permission.

(i) Forming the first disulphide bonds

Reduced BPTI is largely unfolded (Goldenberg & Zhang 1993), although there are local elements of non-random conformation that have small effects on the folding process (Kemmink & Creighton 1993). Being unfolded, the protein forms first disulphide bonds between any of the 15 possible pairs of the six cysteine residues on an almost random basis, varying only approximately by the distance between the pair of residues (Darby & Creighton 1993). Formation of the·

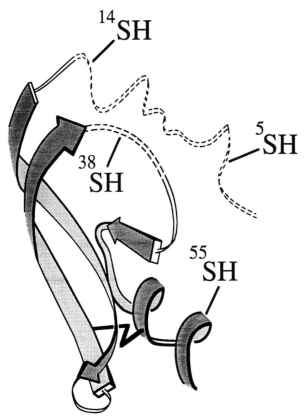

Figure 2. The partly folded conformation of intermediate (30–51) of the BPTI disulphide folding pathway. The portion of the polypeptide chain that is folded into a stable, native-like conformation is indicated in solid lines. The solid cross link is the 30–51 disulphide bond; it is linking the major α-helix of the protein to the β-sheet, which comprise the major hydrophobic core of the native conformation. The portions indicated by broken lines are unfolded or very flexible. The positions of the free cysteine residues are indicated by 'SH'. Taken from van Mierlo *et al.* (1993); used with permission.

three native disulphide bonds directly accounts for only about 5% of the total observed rate.

The resulting one-disulphide intermediates are in relatively rapid equilibrium by intramolecular thiol-disulphide interchange and accumulate according to their relative free energies. They are largely unfolded (Darby *et al.* 1992), except for the quasi-native $(5–55)_N$ and the partly folded (30–51). Intermediate $(5–55)_N$ accumulates to substantial levels at neutral pH, but not at pH 8.7, because the thiol groups of Cys30 and Cys51 are buried in its quasi-native conformation. The kinetically crucial intermediate (30–51) predominates at both pH values and has a partly folded conformation in which about two thirds of the residues have native-like conformations and the remainder are unfolded or very flexible (see figure 2).

(ii) Forming the second disulphide bonds

The tendency of intermediate $(5–55)_N$ to adopt a quasi-native conformation allows it to form the 14–38 disulphide bond rapidly. This is because these two cysteine residues are held in proximity on the surface of the protein, where they can readily react with disulphide reagents and with each other. The resulting $(5–55,14–38)_N$ has an even more stable quasi-native conformation (States *et al.* 1984; Eigenbrot *et al.* 1992)

and completes refolding most readily by first reducing the 14–38 disulphide bond (Creighton & Goldenberg 1984).

The predominance of the (30–51) intermediate is important for increasing the rate of formation of the second disulphide bonds, as the observed rate is 12-fold lower in the presence of 8 M urea, where all the one-disulphide intermediates are unfolded and populated to similar extents (Creighton 1977 *b*). The partly folded conformation of intermediate (30–51) accounts for its tendency to form any of the three possible disulphide bonds between the three cysteine residues that are in the unfolded or flexible regions: Cys5, Cys14 and Cys38 (see figure 2). Of the resulting two-disulphide intermediates, (30–51,5–14), (30–51,5–38) and $(30–51,14–38)_N$, the first two have non-native second disulphide bonds and conformations like that of (30–51), but with the flexible parts of the polypeptide chain tethered by the second disulphide bond (van Mierlo *et al.* 1994). The third intermediate has a quasi-native conformation of virtually all the residues, even in the absence of the 5–55 disulphide bond linking the ends of the polypeptide chain (van Mierlo *et al.* 1991 *a*).

These three two-disulphide intermediates are normally in relatively rapid equilibrium by intramolecular disulphide rearrangements. They do not readily form third disulphide bonds, and they complete refolding most rapidly under usual experimental conditions by rearranging to the native-like N_{SH}^{SH}. This is the slowest intramolecular transition on the normal pathway, with an apparent half-time of about 140 s. It occurs directly, in a single step and at similar rates in both (30–51,5–14) and (30–51,5–38). Intermediate $(30–51,14–38)_N$ probably rearranges via these particular non-native disulphide intermediates, although the more rapid equilibration between these intermediates precludes direct demonstration of this; it also rearranges to $(5–55,14–38)_N$ (Creighton & Goldenberg 1984).

The importance of the pathway via the non-native intermediates is demonstrated by the rapid rates of folding to N_{SH}^{SH} in the absence of just Cys14 or Cys38. This had been demonstrated with chemically blocked thiol groups (Creighton 1977 *a*), but virtually identical results have been obtained with replacement of each of the cysteine residues by serines (Darby *et al.* 1995). One folds through (30–51,5–14), the other through (30–51, 5–38). The rate through this pathway is the same as that observed with normal BPTI. These non-native two-disulphide intermediates are the most productive intermediates preceding N_{SH}^{SH}; in their absence, as when both Cys14 and Cys38 thiol groups are absent, folding is slowed remarkably (Creighton 1977 *a*; Goldenberg 1988). In contrast, the absence of the quasi-native $(5–55,14–38)_N$ and $(30–51,14–38)_N$ does not slow the folding process, indicating that they are not productive intermediates. They accumulate to substantial levels during folding of normal BPTI because of their relative stabilities and because they are blocked in forming the final native disulphide bonds.

Forming the 5–55 or 30–51 disulphide bonds is slow when the stable native conformation will result; this is most evident in (30–51), $(30–51,14–38)_N$, and (5–55,

$14-38)_N$. It does not result from inaccessibility of the thiol groups of the relevant intermediates (Creighton 1981). The reason for the kinetic blockage is apparent upon considering the reverse process, reduction of these disulphide bonds (Creighton 1978). Both the 30–51 and 5–55 disulphide bonds are buried in native BPTI and in N_{SH}^{SH}, and consequently both are reduced directly only very slowly; the transition states for both steps have very high free energies. All the steps in the BPTI pathway are observed to exhibit microscopic reversibility, so the same transition state is encountered in both directions for each step. Therefore, forming the 30–51 or 5–55 disulphide bonds last, to generate three-disulphide native BPTI or N_{SH}^{SH}, will encounter the same high energy transition states, and each step will be correspondingly slow. Consequently, instead of forming the 5–55 disulphide intermediate directly, (30–51) much more rapidly forms the 5–14, 5–38 and 14–38 disulphide bonds and intermediate $(30-51,14-38)_N$ tends to rearrange intramolecularly to N_{SH}^{SH}. A disulphide bond will always be formed slowly if, once formed, it is buried in a stable folded conformation. This is a general phenomenon and is observed in the other protein disulphide folding transitions that have been studied (see below).

The disulphide rearrangements occur in BPTI because the energetic barrier to forming the disulphide bonds that will be buried is overcome in a solely intramolecular transition, rather than in a bimolecular reaction involving a disulphide reagent to form the disulphide bond directly. In fact, the free energy barriers to the intra- and intermolecular reactions are similar in magnitude under normal conditions, suggesting that the same conformational transitions are occurring in each, and that both are likely to involve at least some unfolding of the conformations present (Creighton 1978; Mendoza *et al.* 1994). The rearrangement pathway is not a consequence of the occurrence of the quasi-native species, as it predominates even when they are not populated (Zhang & Goldenberg 1993).

(iii) *Forming the third disulphide bond*

The native-like $(30-51,5-55)_N$, or N_{SH}^{SH}, produced by the disulphide rearrangements has the Cys14 and Cys38 thiol groups in proximity and on the surface of the protein. Consequently, it can rapidly form the third, 14–38 disulphide bond, to complete refolding.

Ribonuclease A (RNase A) RNase A is the classic subject of protein folding (Anfinsen 1973). Nevertheless, its disulphide folding pathway (see Figure 1*b*) has been elucidated only in outline, due to its complexity (Creighton 1977*c*, 1979; Wearne & Creighton 1988; Talluri *et al.* 1994). Disulphide formation between the eight cysteine residues of reduced RNase A is initially random, as in BPTI, but none of the disulphide bonds substantially stabilize non-random conformations of the polypeptide chain. The initial intermediates are largely unfolded, none predominate, and further disulphide formation remains statistical in nature. The conformational tendencies of the RNase A polypeptide

chain appear to be particularly weak; as a consequence, the disulphide intermediates are unstable. These intermediates become increasingly unstable with further disulphide formation, which tends to be slowed, rather than increased, by any disulphide bonds present.

Formation of correctly refolded RNase A is relatively slow. The rate-limiting intramolecular step is formation of the fourth disulphide bond in one or more intermediates with three native disulphide bonds, which undoubtedly adopt the native-like conformation. Two such quasi-native species have been identified: one lacks the 40–95 disulphide bond (Creighton 1980), the other the 65–72 disulphide bond (Talluri *et al.* 1994). The former appeared not to be a productive intermediate, whereas the latter may be (Talluri *et al.* 1994). All four disulphide bonds in native RNase A are buried and inaccessible, and so making any one of them to form all those species with either three or four native disulphide bonds is slow. There is no equivalent of N_{SH}^{SH} of BPTI. Which particular disulphide bonds are reduced first in native RNase A, and formed last upon refolding, depends upon their detailed accessibilities and reactivities in the native conformation, and not folding *per se*.

Ribonuclease T_1 (RNase T_1) Reduced RNase T_1 is unfolded at low salt concentrations, but adopts a quasi-native conformation at high concentrations (Mücke & Schmid 1994). Under conditions where it is unfolded, reduced RNase T_1 forms disulphide bonds readily between any pair of the nearby Cys2, Cys6 and Cys10 and the protein remains unfolded (see figure 1*c*). Forming a disulphide bond between any of these residues and the much more distant Cys103 is much slower, as expected for a random polypeptide chain. Forming the native 6–103 disulphide bond is especially slow, as in this case the protein will adopt the native conformation, in which this disulphide bond will be buried. Instead of being formed directly, the 6–103 disulphide bond is formed most rapidly under the usual experimental conditions by intramolecular rearrangement of the other one-disulphide intermediates (Pace & Creighton 1986).

The $(6-103)_N$ intermediate is comparable to N_{SH}^{SH} of BPTI (see figure 1) because in its native-like conformation the Cys2 and Cys10 residues are proximate and accessible, so that they can readily form the second native disulphide bond and complete refolding.

α-Lactalbumin (αLA) αLA differs from the previous proteins in that it tends to adopt a molten globule type of conformation, and it binds specifically one Ca^{2+} ion (Kuwajima 1989). The molten globule conformation is stabilized only slightly by any disulphide bonds between its eight cysteine residues (Ewbank & Creighton 1991), so reduced αLA adopts this compact conformation about 30% of the time (Ikeguchi & Sugai 1989; Ewbank & Creighton 1993*b*). Otherwise, reduced αLA is unfolded, and it initially forms disulphide bonds randomly, which interconvert readily (see figure 1*d*). The compactness of the molten globule conformation increases the frequency of disulphide formation between cysteine residues distant in the polypeptide chain (Ewbank & Creighton 1993*a*). None of the reduced and one-disulphide αLA species bind Ca²⁺ significantly, so the presence of the ion makes little difference at this stage.

At the two-disulphide stage, an intermediate with two native disulphide bonds (61–77,73–91) is populated significantly in the presence of high concentrations of Ca²⁺ because it can adopt a partly native-like conformation with the Ca²⁺-binding site. The remainder of the polypeptide chain is disordered but compact, and can readily form non-native disulphide bonds between the four free cysteine residues.

In the absence of Ca²⁺, the three-disulphide intermediates have nearly random disulphide bonds, although biased by the presence of the 80 % molten globule conformation (Creighton & Ewbank 1994). One of these species has three native disulphide bonds (28–111,61–77,73–91) and can tightly bind Ca²⁺ when it adopts a quasi-native conformation (Kuwajima *et al.* 1990; Ewbank & Creighton 1993*a*, *b*). The Cys6 and Cys120 cysteine residues are in proximity and readily form the fourth disulphide bond, even though it is strained and super-reactive. The precise pathway followed by αLA in forming and breaking its four disulphide bonds depends upon the Ca²⁺ concentration.

4. LESSONS FROM THE DISULPHIDE FOLDING PATHWAYS

The protein folding process is accelerated if only a few productive intermediates predominate. This is illustrated most clearly with BPTI, as its rate of folding could be measured in 8 M urea, where initial intermediates are unfolded, but the final N_{SH}^{SH} and native protein are not; the rate was decreased 14-fold (Creighton 1977*b*). This effect is relatively modest because just crosslinking the polypeptide chain with disulphide bonds limits the conformational flexibility markedly; consequently, proteins such as RNase A can refold on a reasonable timescale without any productive intermediates predominating. However, completion of disulphide folding can be slowed if the conformations adopted by the intermediates are too stable. The BPTI pathway demonstrates the advantages of having an intermediate (30–51) with a conformation that is sufficiently stable to cause it to predominate, yet sufficiently flexible to permit it to complete folding (see figure 2). It is also advantageous to have a disulphide bond on the surface, such as 14–38 of BPTI, which can be formed rapidly in the final step, after its cysteine residues have participated in intramolecular disulphide rearrangements. *In vivo*, of course, all of these steps are catalysed (Creighton *et al.* 1994).

The disulphide intermediates demonstrate how a folded conformation is stabilized by the effect of multiple interactions stabilizing each other. Whatever conformation is stabilized by a disulphide bond in turn stabilizes that disulphide bond to the same extent (Creighton 1986). This fundamental relation is not specific to disulphide bonds but applies to all other interactions that stabilize folded conformations; it is

probably the key to understanding protein folding and stability. A major challenge then, is to understand why the proteins described here adopt such different conformations in their disulphide intermediates.

Disulphide folding is influenced by the covalent nature of the disulphide bond, its strict geometric requirements, and its formation and breakage only by thiol-disulphide exchange – frequently with an external reagent – and this makes accessibility of thiols and disulphides to the solvent of great importance. These considerations are not so severe with other types of interactions, but they should also apply (although to a lesser extent) to hydrogen bonding: other things being equal, a protein hydrogen bond will be made and broken more rapidly by interchange, perhaps with the aqueous solvent, than in isolation (Creighton 1978).

5. REFERENCES

Anfinsen, C.B. 1973 Principles that govern the folding of protein chains. *Science, Wash.* **181**, 223–230.

Creighton, T.E. 1977*a* Conformational restrictions on the pathway of folding and unfolding of the pancreatic trypsin inhibitor. *J molec. Biol.* **113**, 275–293.

Creighton, T.E. 1977*b* Effects of urea and guanidine HCl on the folding and unfolding of pancreatic trypsin inhibitor. *J. molec. Biol.* **113**, 313–328.

Creighton, T.E. 1977*c* Kinetics of refolding of reduced ribonuclease. *J. molec. Biol.* **113**, 329–341.

Creighton, T.E. 1978 Experimental studies of protein folding and unfolding. *Prog. Biophys. molec. Biol.* **33**, 231–297.

Creighton, T.E. 1979 Intermediates in the refolding of reduced ribonuclease A. *J. molec. Biol.* **129**, 411–431.

Creighton, T.E. 1980 A three-disulphide intermediate in refolding of reduced ribonuclease A with a folded conformation. *FEBS Lett.* **118**, 283–288.

Creighton, T.E. 1981 Accessibilities and reactivities of cysteine thiols during refolding of reduced bovine pancreatic trypsin inhibitor. *J. molec. Biol.* **151**, 211–213.

Creighton, T.E. 1986 Disulfide bonds as probes of protein folding pathways. *Meth. Enzymol.* **131**, 83–106.

Creighton, T.E. 1988 Toward a better understanding of protein folding pathways. *Proc. natn. Acad. Sci. U.S.A.* **85**, 5082–5086.

Creighton, T.E. 1990 Protein folding. *Biochem. J.* **270**, 1–16.

Creighton, T.E. 1992*a* The disulphide folding pathway of BPTI. *Science, Wash.* **256**, 111–112.

Creighton, T.E. 1992*b* Protein folding pathways determined using disulphide bonds. *BioEssays* **14**, 195–199.

Creighton, T.E. 1992*c* Folding pathways determined using disulfide bonds. In *Protein folding* (ed. T. E. Creighton), pp. 301–351. New York: W.H. Freeman.

Creighton, T.E. 1994 The energetic ups and downs of protein folding. *Nature struct. Biol.* **1**, 135–138.

Creighton, T.E. & Ewbank, J.J. 1994 Disulfide-rearranged molten globule state of α-lactalbumin. *Biochemistry* **33**, 1534–1538.

Creighton, T.E. & Goldenberg, D.P. 1984 Kinetic role of a meta-stable native-like two-disulphide species in the folding transition of bovine pancreatic trypsin inhibitor. *J. molec. Biol.* **179**, 497–526.

Creighton, T.E., Zapun, A. & Darby, N.J. 1995 Mechanisms and catalysis of disulphide formation in proteins. *Trends Biotechnol.* **13**, 18–23.

Darby, N.J. & Creighton, T.E. 1993 Dissecting the disulphide-coupled folding pathway of bovine pancreatic trypsin inhibitor. Forming the first disulphide bonds in analogues of the reduced protein. *J. molec. Biol.* **232**, 873–896.

Darby, N.J., van Mierlo, C.P.M. & Creighton, T.E. 1991 The 5–55 single disulphide intermediate in the folding of bovine pancreatic trypsin inhibitor. *FEBS Lett.* **279**, 61–64.

Darby, N.J., van Mierlo, C.P.M., Scott, G.H.E., Neuhaus, D. & Creighton, T.E. 1992 Kinetic roles and conformational properties of the non-native two-disulphide intermediates in the refolding of bovine pancreatic trypsin inhibitor. *J. molec. Biol.* **224**, 905–911.

Darby, N.J., van Mierlo, C.P.M. & Creighton, T.E. 1993 BPTI as a model protein to understand protein folding pathways. In *Innovations on proteases and their inhibitors*. (ed. F. X. Avilés), pp. 391–406. Berlin: Walter de Gruyter.

Darby, N.J., Morin, P.E., Talbo, G. & Creighton, T.E. 1995 Refolding of bovine pancreatic trypsin inhibitor (BPTI) via non-native disulphide intermediates. *J. molec. Biol.* (In the press).

Eigenbrot, C., Randal, M. & Kossiakoff, A.A. 1992 Structural effects induced by mutagenesis affected by crystal packing factors: the structure of a 30–51 disulphide mutant of basic pancreatic trypsin inhibitor. *Proteins struct. funct. Genet.* **14**, 75–87.

Ewbank, J.J. & Creighton, T.E. 1991 The molten globule protein conformation probed by disulphide bonds. *Nature, Lond.* **350**, 518–520.

Ewbank, J.J. & Creighton, T.E. 1993a Pathway of disulfide-coupled unfolding and refolding of bovine α-lactalbumin. *Biochemistry* **32**, 3677–3693.

Ewbank, J.J. & Creighton, T.E. 1993b Structural characterization of the disulfide folding intermediates of bovine α-lactalbumin. *Biochemistry* **32**, 3694–3707.

Goldenberg, D.P. 1988 Kinetic analysis of the folding and unfolding of a mutant form of bovine pancreatic trypsin inhibitor lacking the cysteine-14 and -38 thiols. *Biochemistry* **27**, 2481–2489.

Goldenberg, D.P. & Zhang, J.-X. 1993 Small effects of amino acid replacements on the reduced and unfolded state of pancreatic trypsin inhibitor. *Proteins struct. funct. Genet.* **15**, 322–329.

Hollecker, M. & Creighton, T.E. 1983 Evolutionary conservation and variation of protein folding pathways. Two protease inhibitor homologues from black mamba venom. *J. molec. Biol.* **168**, 409–437.

Hurle, M.R., Eads, C.D., Pearlman, D.A., Seibel, G.L., Thomason, J., Kosen, P.A., Kollman, P., Anderson, S. & Kuntz, I.D. 1992 Comparison of solution structures of mutant bovine pancreatic trypsin inhibitor proteins using two-dimensional nuclear magnetic resonance. *Protein Sci.* **1**, 91–106.

Ikeguchi, M. & Sugai, S. 1989 Contribution of disulfide bonds to stability of the folding intermediate of α-lactalbumin. *Int. J. Peptide Protein Res.* **33**, 289–297.

Kemmink, J. & Creighton, T.E. 1993 Local conformations of peptides representing the entire sequence of bovine pancreatic trypsin inhibitor (BPTI) and their roles in folding. *J. molec. Biol.* **234**, 861–878.

Kosen, P.A., Creighton, T.E. & Blout, E.R. 1983 Circular dichroism spectroscopy of the intermediates that precede the rate-limiting step of the refolding pathway of bovine pancreatic trypsin inhibitor. Relationship of conformation and the refolding pathway. *Biochemistry* **22**, 2433–2440.

Kosen, P.A., Marks, C.B., Falick, A.M., Anderson, S. & Kuntz, I.D. 1992 Disulfide bond-coupled folding of bovine pancreatic trypsin inhibitor derivatives missing one or two disulfide bonds. *Biochemistry* **3**, 5705–5717.

Kuwajima, K. 1989 The molten globule as a clue for understanding the folding and cooperativity of globular-protein structure. *Proteins struct. funct. Genet.* **6**, 87–103.

Kuwajima, K., Ikeguchi, M., Sugawara, T., Hiraoka, Y. & Sugai, S. 1990 Kinetics of disulfide bond reduction in α-lactalbumin by dithiothreitol and molecular basis of superreactivity of the Cys6-Cys120 disulfide bond. *Biochemistry* **29**, 8240–8249.

Mendoza, J.A., Jarstfer, M.B. & Goldenberg, D.P. 1994 Effects of amino acid replacements on the reductive unfolding kinetics of pancreatic trypsin inhibitor. *Biochemistry* **33**, 1143–1148.

Mücke, M. & Schmid, F.X. 1994 Intact disulfide bonds decelerate the folding of ribonuclease T1. *J. molec. Biol.* **239**, 713–725.

Pace, C.N. & Creighton, T.E. 1986 The disulphide folding pathway of ribonuclease T_1. *J. molec. Biol.* **188**, 477–486.

States, D.J., Dobson, C.M., Karplus, M. & Creighton, T.E. 1984 A new two-disulphide intermediate in the refolding of reduced bovine pancreatic trypsin inhibitor. *J. molec. Biol.* **174**, 411–418.

States, D.J., Creighton, T.E., Dobson, C.M. & Karplus, M. 1987 Conformations of intermediates in the folding of the pancreatic trypsin inhibitor. *J. molec. Biol.* **195**, 731–739.

Talluri, S., Rothwarf, D.M. & Scheraga, H.A. 1994 Structural characterization of a three-disulfide intermediate of ribonuclease A involved in both the folding and unfolding pathways. *Biochemistry* **33**, 10437–10449.

van Mierlo, C.P.M., Darby, N.J., Neuhaus, D. & Creighton, T.E. 1991a The 14-38,30–51 double-disulphide intermediate in folding of bovine pancreatic trypsin inhibitor: a two-dimensional ¹H NMR study. *J. molec. Biol.* **222**, 353–371.

van Mierlo, C.P.M., Darby, N.J., Neuhaus, D. & Creighton, T.E. 1991b Two-dimensional ¹H NMR study of the (5–55) single-disulphide folding intermediate of bovine pancreatic trypsin inhibitor. *J. molec. Biol.* **222**, 373–390.

van Mierlo, C.P.M., Darby, N.J. & Creighton, T.E. 1992 The partially folded conformation of the Cys30-Cys51 intermediate in the disulphide folding pathway of bovine pancreatic trypsin inhibitor. *Proc. natn. Acad. Sci. U.S.A.* **89**, 6775–6779.

van Mierlo, C.P.M., Darby, N.J., Keeler, J., Neuhaus, D. & Creighton, T.E. 1993 Partially folded conformation of the (30–51) intermediate in the disulphide folding pathway of bovine pancreatic trypsin inhibitor. ¹H and ¹⁵N Resonance assignments and determination of backbone dynamics from ¹⁵N relaxation measurements. *J. molec. Biol.* **229**, 1125–1146.

van Mierlo, C.P.M., Kemmink, J., Neuhaus, D., Darby, N.J. & Creighton, T.E. 1994 ¹H-NMR analysis of the conformational properties of the partially-folded non-native two-disulphide bonded intermediates (30–51,5–14) and (30–51,5–38) in the disulphide folding pathway of bovine pancreatic trypsin inhibitor. *J. molec. Biol.* **235**, 1044–1061.

Wearne, S.J. & Creighton, T.E. 1988 Further experimental studies of the disulfide folding transition of ribonuclease A. *Proteins struct. funct. Genet.* **4**, 251–261.

Weissman, J.S. & Kim, P.S. 1991 Reexamination of the folding of BPTI: predominance of native intermediates. *Science, Wash.* **253**, 1386–1393.

Weissman, J.S. & Kim, P.S. 1992 Kinetic role of the non-native species in the folding of bovine pancreatic trypsin inhibitor. *Proc. natn. Acad. Sci. U.S.A.* **89**, 9900–9904.

Zhang, J.X. & Goldenberg, D.P. 1993 An amino acid replacement that eliminates kinetic traps in the BPTI folding pathway. *Biochemistry* **32**, 14075–14081.

Mapping the structures of transition states and intermediates in folding: delineation of pathways at high resolution

A. R. FERSHT

Cambridge Centre for Protein Engineering, Department of Chemistry, University of Cambridge, Lensfield Road, Cambridge CB2 1EW, U.K.

SUMMARY

The structures of all the intermediates and transition states, from the unfolded state to the native structure, are being determined at the level of individual residues in the folding pathways of barnase and chymotrypsin inhibitor 2 (CI2), using a combination of protein engineering and nuclear magnetic resonance methods. Barnase appears to refold according to a classical framework model in which elements of secondary structure are flickeringly present in the denatured state, consolidate as the reaction proceeds and, when nearly fully formed, dock in the rate-determining step. Unlike barnase, CI2 folds without a kinetically significant folding intermediate. The transition state for its formation has no fully formed elements of secondary structure, and the transition state is like an expanded form of the native structure. CI2 probably represents the folding of an individual domain in a larger protein, whereas barnase represents the folding of a multi-domain protein. The protein engineering methods are being extended to map the pathway in the presence of molecular chaperones. There are parallels between the folding of barnase when bound to GroEL and in solution.

1. INTRODUCTION

(a) Pathways of protein folding

Levinthal pointed out some 25 years ago that, because of the astronomical number of conformations available to an unfolded polypeptide chain, protein folding could not occur by a random, unbiased search of all conformational space (Levinthal 1968). This has led to the idea that there must be defined pathways of folding by which only a small fraction of conformational space is searched. These mechanisms may be divided up into two basic classes (figure 1). The first involves secondary structure being formed before tertiary structure: 'framework models' (Ptitsyn 1973, 1991). It is postulated that small segments of native-like secondary structure form locally and rapidly. These then lead to the formation of tertiary structure. One sub-division of the mechanism is the 'diffusion–collision' model (Karplus & Weaver 1976, 1994): the elements of secondary structure diffuse, collide and coalesce. The second sub-division is 'propagation' (Wetlaufer 1973): an element of secondary structure acts as a nucleation point and structure spreads out from this. The second class of mechanisms involves a general hydrophobic collapse of the protein, from which state rearrangement occurs, leading to the formation of secondary structure.

(b) The strategy for solving pathways

There is a simple strategy that has been used by chemists and biochemists for elucidating mechanisms of reactions and pathways that can be directly translated into analysing the pathway of protein folding. The pathway of folding will be defined when the structures of all intermediates and transition states have been characterized from the initial unfolded state to the final folded state. Until recently, the experimental means of tackling the problem of protein folding at high resolution have not been available. There are now two procedures that can characterize the necessary structures at the level of individual residues and even atoms. Nuclear magnetic resonance (NMR) is unparalleled as a spectroscopic technique in the level of detail obtainable for structure in solution. The power of this method can be applied to analysing stable states, including the unfolded state, and also to be used for studying transients by H–D exchange. For highly defined structures, information is given about backbone and side chains. For less characterized structures and H–D exchange, information tends to be more about the backbone and secondary structure.

The only way of analysing transition states is by kinetics, and the only way of analysing detail at individual residues is by kinetic measurements on mutant enzymes: the 'protein engineering approach' (Matouschek et al. 1989, 1990; Arcus et al. 1994). The protein engineering method may also be used to characterize intermediates. This procedure has been described in detail (Fersht et al. 1992) and recently in a more general review (Fersht 1993). The principle of the method is very simple. A suitable side chain in the protein is mutated, generally to a smaller one or to that of alanine, and the change in stability of the protein is measured. The effects of mutation on the stability of intermediates and transition states are also measured. If, for example, a transition state is destabilized by

Phil. Trans. R. Soc. Lond. B (1995) **348**, 11–15
Printed in Great Britain

11

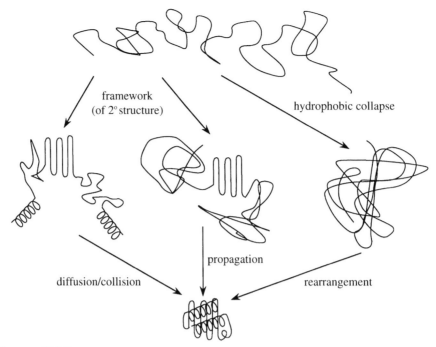

framework
(of 2° structure)

hydrophobic collapse

propagation

diffusion/collision

rearrangement

Figure 1. Pathways of folding.

exactly the same amount as the folded state on mutation, then the structure at the site of mutation is most likely the full native structure. If, on the other hand, the mutation does not alter the stability of the transition state at all, then the protein must be fully unfolded at the site of the mutation. These principles may be converted into quantitative terms by defining a parameter ϕ which is the ratio of change in energy of a particular state to the change in energy of the native state on mutation. $\phi = 0$ means structure is not formed; $\phi = 1$ means structure is fully formed.

2. DISSECTION OF THE FOLDING PATHWAY OF BARNASE

Barnase is a small ribonuclease of 110 amino acid residues that is an almost perfect paradigm for studying the folding of a small protein that folds via a distinct folding intermediate. Importantly, it has no disulphide

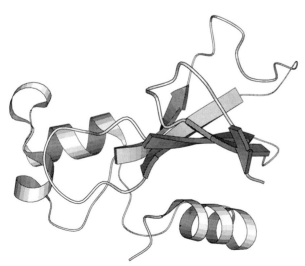

Figure 2. Structure of barnase. The major α-helix is at the bottom.

cross-links that prevent its unfolding fully and so it can be used to analyse events that are presumed to follow from peptide biosynthesis onwards. The structure (figure 2) is characterized by a major helix (residues 6–18) docking on to a five-stranded anti-parallel β-sheet. The hydrophobic side chains from each element of secondary structure interdigitate to form the hydrophobic core. We now work backwards from the fully folded structure to the unfolded state showing how those elements of structure persist or become embryonic as the structure unfolds.

(a) The unfolding transition state

The protein engineering method has been applied systematically around the enzyme to map out the structure of the transition state for unfolding and the subsequent folding intermediate. Most ϕ values are either 1 or 0, indicating that structure is close to being fully formed or fully unformed in the transition state. The major helix is formed from residues 8–18. The sheet is basically formed but is weaker at the edges. The hydrophobic core is in the process of being formed. Individual details have been summarized by Fersht (1993).

(b) The folding intermediate

This is similar to the transition state but slightly more weakened at the edges of the sheet and throughout the core.

(c) The unfolded state

One could be excused from thinking that the unfolded state is the least interesting on the pathway. But, in many ways it is the most interesting because it could contain the clues as to whether framework or collapsed models hold. Until recently, unfolded states could not be studied at high resolution: NMR spectroscopy was obscured by poor dispersion of the signals. However, the laboratory of Wüthrich has introduced the application of NMR magnetization transfer to assign an unfolded state (Neri *et al.* 1992; Wüthrich 1994), and the laboratory of Fesik (Logan *et al.* 1994) has used multinuclear methods for assigning another protein. We have combined both approaches to assign fully the unfolded state of barnase (Arcus *et al.* 1994). Although the structure has not yet been solved, there is sufficient information from chemical shifts and sequential nuclear overhauser effects (NOEs) to implicate regions of structure. First, there is a tendency for residues 6–18 to be in their helical formation. Second, the centre of the β-sheet has a collapsed structure. This appears to be non-native but can probably rearrange to native-like structure easily.

(d) Summary of folding pathway

Barnase thus appears to fold by a framework model. Weak structure in the major α-helix and the β-sheet is present in the unfolded state. These consolidate in a folding intermediate, tighten up in the transition state as the hydrophobic core consolidates to give the final folded structure.

3. FOLDING PATHWAY OF THE BARLEY CI2 INHIBITOR

The barley CI2 inhibitor is a small (64 residue) polypeptide that specifically inhibits chymotrypsin and subtilisin. It is an excellent paradigm for studying the folding of a very small protein that folds without there being a kinetically significant intermediate (Jackson &

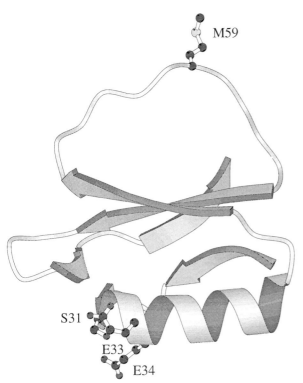

Figure 3. Structure of CI2. The single methionine is illustrated as are the residues at the N-terminal region of the helix which constitute the best-formed region in the transition state for refolding.

Fersht 1991). There is just a single rate-determining state linking the unfolded and folded state. This transition state has been analysed by the protein engineering methods both in the direction of folding and unfolding (Jackson *et al.* 1993; Otzen *et al.* 1994). The structure is found to be the same when measured in both directions. The transition state for the folding of CI2 is quite different in nature from that for barnase. Most of the φ values are fractional, there being no values of 1. The best formed part of the structure is the N-terminal region of the single α-helix. The rest of the protein is like the folded structure that has been expanded by 30%.

(a) Molecular dynamics simulation of the transition state

Li & Daggett (1994) have simulated the unfolding of CI2 using molecular dynamics. The picture they obtained is in remarkable agreement with that found from the protein engineering method. Their calculations flesh out in structural terms what the fractional φ values mean. Their structure shows a molecule with a distorted helix, the core coming apart and the sheet disintegrating. Caflisch & Karplus (1994) have simulated the transition state for the unfolding of barnase, again in agreement with experiments.

(b) Fragments of CI2

Peptide fragments of proteins provide a convenient and attractive way of analysing some characteristics of the unfolded states. Small fragments are more tractable to analysis by NMR, and can be studied in simple aqueous solution in water in the absence of denaturants. They also provide the means of looking for local effects in structure since residues far removed in sequence may be physically removed in the fragment. CI2 may be conveniently cleaved into two fragments using CNBr on the single methionine residue at position 59. A whole family of fragments of CI2 has been generated by cleaving all mutants made so far in this laboratory at their unique methionine 59 (Prat Gay & Fersht 1994; Prat Gay *et al.* 1994*a, b*). These two fragments recombine to give native-like structure. NMR studies on the complex show that it is almost identical to the native structure apart from the loop in which cleavage has occurred (B. Davies & A. R. Fersht, unpublished). The individual fragments have collapsed structure, but have non-native hydrophobic clusters being formed. The protein engineering method has been used to analyse the transition state for the association of the two fragments (Prat Gay *et al.* 1994*b*). It is remarkably like the transition state for the formation of an intact protein. Thus, two collapsed non-native structures can associate to give a folded protein via a transition state that is similar to that of the formation of the native protein.

(c) Summary of folding pathway of CI2

The folding pathway of CI2 thus appears to involve the rearrangement of a non-active collapsed structure to form the folded structure by a single transition state.

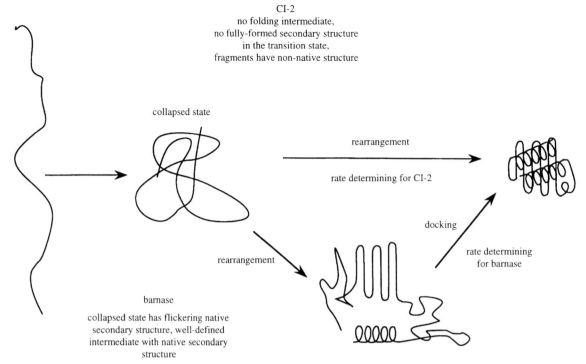

Figure 4. Unified scheme for the folding of the multi-domain (multi-module) protein barnase and the single-domain (single module) CI2.

3. UNIFIED SCHEME FOR THE FOLDING OF CI2 AND BARNASE

Barnase, although being small, does appear to have sub-structures that Go calls modules (Yanagawa *et al.* 1993), i.e. regions that make most of their interactions within themselves. CI2, on the other hand, appears to be like a single module, there being no substructures. We suggest that the folding of CI2 is representative of the folding of single modules of structure, whereas barnase represents the folding of larger structures that contain separate modules. Thus, we postulate, as in figure 4, that the folding pathway of CI2 involves the rearrangement of its collapsed state to that of the fully folded module (Otzen *et al.* 1994). Barnase first forms a state that has collection of regions of secondary structure in some of its individual modules which then coalesce in the rate-determining step to form the final folded structure. Thus, barnase folds by a clear framework model because it is large enough to do so. However, the framework does have an initial collapse step.

4. FOLDING *IN VITRO* VERSUS FOLDING *IN VIVO*

Protein folding *in vivo* is now known to involve a series of accessory proteins, which will be discussed by Hartl, Sigler, Tokatlidis *et al.* and Jaenicke in this volume. Members of the hsp70 class of protein (DnaK) coat the nascent polypeptide chain as it is synthesized on the ribosome. The polypeptide chain is then released under the influence of ATP and other proteins to be transferred to GroE whose structure is detailed by Sigler (this volume). This very large protein consists of two doughnut-shaped rings of 7-mers stacked one upon

the other to give a large central cavity. There is currently controversy as to where the polypeptide binds and as to whether or not the polypeptide can fold while being bound to GroE or it has to be released into solution to do so.

We can tackle the folding of barnase in the presence of GroE using the protein engineering method. The use of mutants is one of the most powerful procedures for studying proteins *in vivo* and so our refined method adapts readily to provide important information about the pathway of folding in the presence of accessory proteins.

(a) Barnase folds while binding to GroE

Unfolded barnase is found to bind to GroE: when the two are mixed together, the rate of refolding of barnase is slowed down as measured by the regain in its catalytic activity (Gray & Fersht 1993). Further, the rate constant for the regain of activity levels off to a constant value in the presence of saturating concentration of GroE. This means that a folding event must occur while the peptide is bound to GroE because otherwise, by the principle of mass action, any unfolded barnase that is released would be mopped up by saturating concentrations of GroE. Further, the rate of refolding of barnase in the presence of saturating GroE is sensitive to mutations in barnase that are known to affect the rate of folding (Gray *et al.* 1993).

(b) Folding when bound to GroE parallels folding in solution

We have measured the rate constant for the refolding of barnase mutants in solution for the step of the intermediate progressing to the folded state. We found that the rate constants for the refolding of the same mutants when bound to GroE are all scaled down by approximately the same factor (Gray *et al.* 1993). This shows that the refolding rate constant in the presence of GroE responds to mutation in the same way as does the transition from intermediate to folded state in solution. This is consistent with the pathway of folding in the presence of GroE being essentially the same as that in solution. The most likely explanation is that barnase refolds within the central cavity of GroE, consistent with the idea of GroE acting as a cage in which molecules can fold without competition from aggregation (Ellis 1994). Although there is evidence that larger proteins require dissociation from GroE for folding to occur (Weissman *et al.* 1994), it is possible that barnase is a model for the folding of small parts of larger molecules in the presence of GroE. Given that a protein as small as barnase can fold within the central cavity, it is thus quite likely that larger proteins could fold by different parts of their structure entering the cavity during various collisions and associations.

Studies on the spontaneous refolding of proteins *in vitro* and in the presence of molecular chaperones are thus converging. It is likely that within the next few years, the pathways of folding will be known at high resolution.

5. REFERENCES

Arcus, V.L., Vuilleumier, S., Freund, S.V., Bycroft, M. & Fersht, A.R. 1994 Towards solving the folding pathway of barnase: the complete backbone ^{13}C, ^{15}N and ^{1}H NMR assignments of its pH-denatured state. *Proc. natn. Acad. Sci. U.S.A.* **91**, 9412–9416.

Caflisch, A. & Karplus, M. 1994 Molecular-dynamics simulation of protein denaturation – solvation of the hydrophobic cores and secondary structure of barnase. *Proc. natn. Acad. Sci. U.S.A.* **91**, 1746–1750.

Ellis, R.J. 1994 Roles of molecular chaperones in protein folding. *Curr. Opin. struct. Biol.* **4**, 117–122.

Fersht, A.R. 1993 Protein folding and stability – the pathway of folding of barnase. *FEBS Lett.* **325**, 5–16.

Fersht, A.R., Matouschek, A. & Serrano, L. 1992 The folding of an enzyme. 1. Theory of protein engineering analysis of stability and pathway of protein folding. *J. molec. Biol.* **224**, 771–782.

Gray, T.E., Eder, J., Bycroft, M., Day, A.G. & Fersht, A.R. 1993 Refolding of barnase mutants and pro-barnase in the presence and absence of GroEL. *EMBO J.* **12**, 4145–4150.

Gray, T.E. & Fersht, A.R. 1993 Refolding of barnase in the presence of GroE. *J. molec. Biol.* **232**, 1197–1207.

Jackson, S.E., elMasry, N. & Fersht, A.R. 1993 Structure of the hydrophobic core in the transition state for folding of chymotrypsin inhibitor-2 – a critical test of the protein engineering method of analysis. *Biochemistry* **32**, 11270–11278.

Jackson, S.E. & Fersht, A.R. 1991 Folding of chymotrypsin inhibitor-2. 1. Evidence for a two-state transition. *Biochemistry* **30**, 10428–10435.

Karplus, M. & Weaver, D.C. 1976 Protein folding dynamics. *Nature, Lond.* **260**, 404–406.

Karplus, M. & Weaver, D.L. 1994 Protein folding dynamics – the diffusion – collision model and experimental data. *Protein Sci.* **3**, 650–668.

Levinthal, C. 1968 Are there pathways for protein folding? *J. chem. Phys.* **85**, 44–45.

Li, A. & Daggett, V. 1994 Characterization of the transition state of protein unfolding using molecular dynamics: chymotrypsin inhibitor 2. *Proc. natn. Acad. Sci. U.S.A.* **91**, 10430–10434.

Logan, T.M., Theriault, Y. & Fesik, S.W. 1994 Structural characterization of the FK506 binding protein unfolded in urea and guanidine hydrochloride. *J. molec. Biol.* **236**, 637–648.

Matouschek, A., Kellis, J.T. Jr, Serrano, L., Bycroft, M. & Fersht, A.R. 1990 Transient folding intermediates characterized by protein engineering. *Nature, Lond.* **346**, 440–445.

Matouschek, A., Kellis, J.T. Jr, Serrano, L. & Fersht, A.R. 1989 Mapping the transition state and pathway of protein folding by protein engineering. *Nature, Lond.* **342**, 122–126.

Neri, D., Billeter, M., Wider, G. & Wuthrich, K. 1992 NMR determination of residual structure in a urea-denatured protein, the 434-repressor. *Science, Wash.* **257**, 1559–1563.

Otzen, D.S.I.L., ElMasry, N., Jackson, S.E. & Fersht, A.R. 1994 The structure of the transition state for the folding/unfolding of the barley chymotrypsin inhibitor 2 and its implications for mechanisms of protein folding. *Proc. natn. Acad. Sci. U.S.A.* **91**, 10422–10425.

Prat Gay, G. de & Fersht, A.R. 1994 Generation of a family of protein fragments for structure-folding studies. I: Folding complementation of two fragments of chymotrypsin inhibitor-2 formed by cleavage at its unique methionine residue. *Biochemistry* **33**, 7957–7963.

Prat Gay, G. de, Ruiz-Sanz, J., Davis, B. & Fersht, A.R. 1994a The structure of the transition state for the association of two fragments of barley chymotrypsin inhibitor 2 to generate native-like protein: Implications for mechanisms of protein folding. *Proc. natn. Acad. Sci. U.S.A.* **91**, 10943–10946.

Prat Gay, G. de, Ruiz-Sanz, J. & Fersht, A.R. 1994b Generation of a family of protein fragments for structure-folding studies. II. Kinetics of association of the two chymotrypsin inhibitor-2 fragments. *Biochemistry* **33**, 7964–7970.

Ptitsyn, O.B. 1973 Stage mechanism of the self-organization of protein molecules. *Dokl. Acad. Nauk. SSR* **210**, 1213–1215.

Ptitsyn, O.B. 1991 How does protein synthesis give rise to the 3D-structure. *FEBS Lett.* **285**, 176–181.

Weissman, J.S., Kashi, Y., Fenton, W.A. & Horwich, A.L. 1994 Groel-mediated protein-folding proceeds by multiple rounds of binding and release of nonnative forms. *Cell* **78**, 693–702.

Wetlaufer, D.B. 1973 Nucleation, rapid folding, and globular intrachain regions in proteins. *Proc. natn. Acad. Sci. U.S.A.* **70**, 697–701.

Wuthrich, K. 1994 NMR assignments as a basis for structural characterization of denatured states of globular proteins. *Curr. Opin. struct. Biol.* **4**, 93–99.

Yanagawa, H., Yoshida, K., Torigoe, C., Park, J.S., Sato, K., Shirai, T. & Go, M. 1993 Protein anatomy – functional roles of barnase module. *J. biol. Chem.* **268**, 5861–5865.

Insights into protein folding using physical techniques: studies of lysozyme and α-lactalbumin

SHEENA E. RADFORD AND CHRISTOPHER M. DOBSON

Oxford Centre for Molecular Sciences, New Chemistry Laboratory, University of Oxford, South Parks Road, Oxford OX1 3QT, U.K.

SUMMARY

Understanding the process of protein folding, during which a disordered polypeptide chain is converted into a compact well-defined structure, is one of the major challenges of modern structural biology. In this article we discuss how a combination of physical techniques can provide a structural description of the events which occur during the folding of a protein. First, we discuss how the rapid kinetic events which take place during *in vitro* folding can be monitored and deciphered in structural terms. Then we consider how more detailed structural descriptions of intermediates may be obtained from NMR studies of stable, partly folded states. Finally, we discuss how these experimental strategies may be extended to relate the findings of *in vitro* studies to the events occurring during folding *in vivo*. The approaches will be illustrated using results primarily from our own studies of the c-type lysozymes and the homologous α-lactalbumins. The conclusions from these studies are also related to those from other systems to highlight their unifying features. On the basis of these results we identify some of the determinants of the events in folding and we speculate on the importance of these in driving folding molecules to their native states.

1. INTRODUCTION

Proteins are synthesized within the cell on ribosomes. Although folding in the cell is a highly complex process involving a cascade of helper proteins called the molecular chaperones, it is clear that the major events in the folding process occur after departure from the ribosome (Ellis 1994; Frydman *et al.* 1994) and perhaps even after release from the chaperones (Weissman *et al.* 1994). For extracellular proteins the process is even more complex because folding must be tightly coupled to translocation (Bychkova & Ptitsyn 1994) and in this case a large part of folding may occur following secretion from the cell. Despite these complexities, many proteins can fold efficiently and correctly in isolation, provided that suitable conditions are found.

Given the improbability that folding could occur in a finite time on a random search basis (Levinthal 1968; Sali *et al.* 1994) and that folding in the cell is not thought to be sterically driven by chaperones, but merely controlled, it is unlikely that the principles behind the folding process *in vivo* and *in vitro* will be fundamentally different (although the two situations may differ substantially in detail given the far from physiological conditions of most experiments *in vitro*). Studies of folding *in vitro*, where physical techniques capable of providing detailed structural information can be used most readily, can provide specific information about the folding process at the molecular level. It is then of great interest to investigate how the various auxiliary factors might moderate or control the process in the light of our understanding of the refolding of proteins in isolation. Efforts to extend the use of physical techniques to enable the achievement of these objectives are increasingly being made (Gray &

Fersht 1993; Jackson *et al.* 1993; Hartl *et al.* 1994; Landry & Gierasch 1994; Robinson *et al.* 1994).

In this article, strategies to provide a structural description of the events in protein folding are discussed with particular reference to one family of proteins, the c-type lysozymes; these are small monomeric proteins with *ca.* 130 residues. The structure of the archetypal family member of this family – the hen protein – is shown in figure 1. There are four α-helices; two located towards the C-terminus and two towards the N-terminus of the polypeptide chain, which together with a short 3_{10} helix make up one 'domain' of the molecule (the α-domain). The other domain, the β-domain, is

Figure 1. Schematic view of the structure of hen lysozyme (Blake *et al.* 1965). The α- and β-domains are shaded white and grey, respectively, and the four disulphide bonds are shown in black. The diagram was drawn using the program Molscript (Kraulis 1991).

Phil. Trans. R. Soc. Lond. B (1995) **348**, 17–25
Printed in Great Britain

17

formed from a triple-stranded antiparallel β-sheet, another short 3_{10} helix, and a long loop. A short antiparallel double-stranded β-sheet links these two domains, as does one of the four disulphide bridges. The disulphide bonds remain intact during the majority of the folding studies discussed here.

2. KINETIC REFOLDING *IN VITRO*

Refolding of most small proteins following dilution, for example, from high concentrations of urea or guanidinium chloride, is often completed within a second or so. Given that NMR spectroscopy is an inherently insensitive and slow technique, the simple view that structural changes taking place during refolding could be monitored by recording a series of spectra as a function of time during refolding, although appealing in principle is, for the majority of cases, impossible in practice. Only in a few cases, for example where proteins fold very slowly because of a need for cis/trans proline isomerism, has direct observation of folding by one-dimensional NMR methods been possible (Koide *et al.* 1993).

Despite these problems, methods have been developed to allow the power of NMR spectroscopy to be applied to the study of the structures of species sampled during folding by adopting an indirect approach. The key to this is that the rate of exchange of amide hydrogens with solvent is dependent on their environment in a protein structure. In the pulse labelling approach (Roder *et al.* 1988; Udgaonkar & Baldwin 1988), this property is utilized to trap deuterons at specific amide sites within regions of structure as they are formed during folding. These sites can be detected and assigned to individual residues by two-dimensional NMR methods by recording spectra of the protein once refolding is complete and making use of residue-specific assignments of the native protein. By this method, it was found that the two structural domains of lysozyme (despite undergoing cooperative unfolding and refolding at equilibrium; Imoto *et al.* 1972) form structure protected against hydrogen exchange at very different rates. In the majority of molecules, the α-domain forms faster than the β-domain (Radford *et al.* 1992). Although within both domains the protection kinetics appear similar (indicating that stabilization of each domain to a protected state occurs cooperatively), protection does not follow a simple exponential path. Further experiments have shown that this is because there are several alternate folding routes (Radford *et al.* 1992; Miranker *et al.* 1993).

To complement these studies we have introduced the idea of monitoring the cooperativity of folding and the populations of folding intermediates by detection of hydrogen exchange labelling by electrospray ionization mass spectrometry (ESI-MS) (Miranker *et al.* 1993). The basis of this method is that the incorporation of deuterons instead of protons within a protein molecule increases its mass, and that ESI-MS enables this to be measured, for a protein the size of lysozyme to a mass resolution better than 1 Da. This approach has enabled us to show that the α-domain can fold to a protected state independently of the β-domain, although the converse is not true, and there is no evidence to suggest that persistent structuring of the β-domain can occur in the absence of a stably structured α-domain. In addition, the multi-exponential protection kinetics arise from the existence of populations of molecules with distinct folding kinetics; some 25% of the population appears able to fold to a native-like state within 10 ms, whereas in the remainder of the molecules this process takes in excess of 300 ms.

These results provide considerable insight into the nature of species formed during the folding process. They rely, however, on hydrogen exchange protection which, because of the manner in which the experiments are performed, can only detect highly stabilized folding intermediates. To gather information about less structured states and to interpret further the structural details of protected states, these methods must be related to other indicators of structure gained through complementary techniques (Dobson *et al.* 1994; Evans & Radford 1994). We have therefore carried out a variety of stopped-flow optical measurements under conditions as similar as possible to the hydrogen exchange measurements (Radford *et al.* 1992; Itzhaki *et al.* 1994). A summary of these experiments and their results is given in table 1. Of particular importance are circular dichroism (CD) experiments in the far ultraviolet (UV) region, which monitors formation of secondary structure, and in the near UV which detects the immobilization of aromatic residues within a close packed structure. In addition, intrinsic fluorescence, fluorescence quenching and binding of the dye 1-anilino naphthalene sulphonic acid (ANS), which are thought to reflect hydrophobic collapse, have been measured. Finally, formation of the functional native protein can be detected through binding of a fluorescently labelled inhibitor that binds to the active site cleft which lies between the two folding domains (Itzhaki *et al.* 1994).

3. SCHEMATIC PATHWAYS OF FOLDING

The acquisition of this array of structural information enables us to propose a schematic folding pathway for lysozyme that is consistent with the currently available experimental evidence (see figure 2). This indicates that within a few ms a compact state of the protein is formed which has a native-like content of secondary structure (Chaffotte *et al.* 1992; Radford *et al.* 1992). Primary evidence for this comes from the very rapid kinetics of formation of native-like ellipticity monitored by far UV CD (Chaffotte *et al.* 1992; Radford *et al.* 1992), from the observation that ANS fluorescence is enhanced most strongly at the earliest measurable time following initiation of folding, and from the resilience of this state to fluorescence quenching by iodide ions (Itzhaki *et al.* 1994).

This state is shown in figure 2 to be heterogeneous; this provides a mechanism for kinetically distinct populations of folding molecules. The ESI-MS data require that such heterogeneity occurs early in folding before the onset of protection (Miranker *et al.* 1993), but refolding experiments from denatured states generated by different means indicates that residual structure in

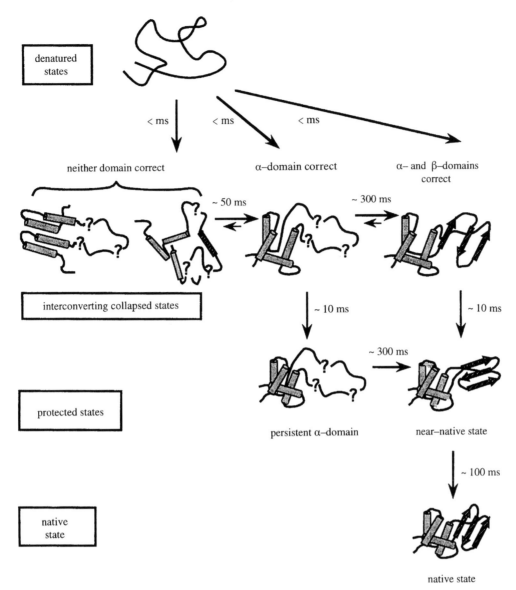

Figure 2. Schematic representation of a possible folding pathway of hen lysozyme deduced from the eight complementary methods listed in table 1. The four native disulphide bonds that remain intact throughout the folding process are not shown.

guanidinium chloride is unlikely to be the origin of this phenomenon (Kotik *et al.* 1995). Rather, we believe it arises from the existence of both correctly folded and misfolded secondary structure or the existence of distinct topological forms resulting from the rapid collapse of the polypeptide chain; recent theoretical predictions provide support for this general idea (Thirumalai & Guo 1995). These forms may be particularly significant in the folding of lysozyme because the polypeptide chain is constrained by the four native disulphide bonds. Studies of the folding of a three disulphide derivative of lysozyme lacking the disulphide bond (Cys 6-Cys 127) which links the N- and C-terminal regions shows, however, that although the presence of this disulphide bond stabilizes the α-domain intermediate, its loss does not appear to change the folding pathway fundamentally (Denton *et al.* 1994; Eyles *et al.* 1994).

At the early stage (< ms) of folding the structure formed is not persistent, indicating that these states are highly dynamic and rapidly fluctuating. Estimates of protection factors for the amide hydrogens in these states indicate that these are less than *ca.* 50. The formation of persistent structure occurs in the next steps along the pathway, and is indicated by protection factors in excess of 500 (Radford *et al.* 1992). Most molecules fold to form persistent structure in the α-domain on a *ca.* 50 ms timescale, whereas the formation of a stably structured β-domain takes significantly longer and is not complete in most molecules until at least 300 ms. On the pathway shown in figure 2, however, a fast folding population of molecules is seen to form persistent structure in both domains on a *ca.* 10 ms timescale. The fact that no such phase is seen in the fluorescent inhibitor binding experiments shows that the active site cleft has not yet formed in these molecules and suggests that another step is needed to bring the domains together to form the fully native state. The timescale of this process is not resolved from that of formation of the native enzyme in the population of slow folding molecules, suggesting that if this step is common to both pathways, it must occur on

Table 1. *Different techniques used to determine the kinetic folding pathway of hen lysozyme*

(All experiments were performed at pH 5.2, 20 °C. The initial denatured state was formed in 6 M guanidinium chloride and refolding was initiated by 11-fold dilution into the refolding buffer. Relevant references to this work are given in the text. HX, hydrogen exchange.)

method	information	result
HX labelling and NMR		
amides	formation of persistent secondary structure	parallel pathways observed: the major folding intermediate is protected in the α-domain
Trp indoles	formation of persistent tertiary structure	hydrophobic core formation precedes α-domain stabilization
HX labelling and ESI-MS	detection of intermediates and folding populations	parallel paths confirmed: independent folding of α-domain, but not the β-domain, demonstrated
far UV CD	formation secondary structure (and possibly some tertiary interactions)	(i) near native secondary structure formed within 2 ms (ii) ellipticity greater than that of the native structure seen after *ca.* 50 ms (iii) slowest phase monitors formation of the ellipticity of the native state
near UV CD	immobilization of aromatic residues	single exponential similar to (iii) above
intrinsic fluorescence	environment of Trp and Tyr residues	(i) change in environment of Trp residues occurs within 2 ms (ii) excess fluorescence quenching in intermediate at *ca.* 50 ms (iii) slow phase monitors recovery of fluorescence to value of native state
fluorescence quenching	solvent accessibility of fluorophore	(i) substantial protection in 2 ms indicates burial of Trp residues (ii) native-like resistance to quenching achieved within *ca.* 100 ms
ANS binding	exposure of hydrophobic surface	2 ms intermediate shows maximum fluorescence enhancement with ANS
inhibitor binding	formation of native state	native state forms in the slowest observable phase; only a single kinetic step is observed at 20 °C

a timescale of about 100 ms. Preliminary results investigating the temperature dependence of folding support this result (A. Matagne, S.E. Radford & C.M. Dobson, unpublished data).

The existence of a schematic pathway is of considerable value in the design of further experimental strategies to test it. Particular areas of interest concern the nature of the collapsed states, the apparent inability of the β-domain to fold to stable structure independently of the α-domain, and the origin of the cooperativity involved in the formation of the structural domains. One strategy we have adopted to explore the former is to probe the development of exchange protection of side chain hydrogens of tryptophan residues (see table 1). Lysozyme has six such residues, four of which are located within the hydrophobic core in the α-domain of the native protein (Blake *et al.* 1965). The kinetics of protection of these hydrogens have been studied in detail, and preliminary data suggest that for several of these protection occurs, at least to a limited extent, before the development of the stable α-domain (Radford *et al.* 1992; C. Morgan, A. Miranker & C.M. Dobson, unpublished data). This, like the ANS binding and fluorescence quenching experiments, suggests that a rudimentary hydrophobic core, at the very least, is formed early during the folding process, from, or within which, the development of persistent secondary structure takes place.

A further approach is to examine the folding of lysozymes from different species. The most important finding from these studies so far is that the refolding of the human variant (which is about 40% identical in sequence to hen lysozyme), while resembling that of the hen protein in outline, differs from it in detail (Hooke *et al.* 1994). Specifically, although refolding studies of the human protein show that the α-domain forms before the β-domain in most molecules, formation of the former domain is not fully cooperative and amides located in the A-, B- and C-terminal 3_{10} helices become protected before those in the C- and D-helices. Thus it seems that this region of structure forms a distinct stable subdomain in the folding process. Although such a subdomain is not observed in the hen protein, its absence suggests that if it is formed it is not stable enough to resist hydrogen-exchange labelling. Although specific site-directed mutants will be needed to probe the molecular origin of such a difference, the observation itself provides evidence that the cooperative but local assembly of regions of secondary structure is a key aspect of folding.

4. THE STRUCTURES OF FOLDING INTERMEDIATES

The combination of the different kinetic techniques described above enables the structural characteristics of the various intermediates to be inferred from these experiments. To test the conclusions from such studies, however, much more detailed structural studies are required. One approach to achieving this is to produce and study stable analogues of the transient intermediates which allow the full potential of NMR in the elucidation of the structure and dynamics of these states to be realized (Baum *et al.* 1989; Hughson *et al.* 1990; Darby *et al.* 1991; Alexandrescu *et al.* 1993). If characteristics such as their optical and hydrogen exchange properties can be correlated with those of their transient counterparts, then the conclusions from detailed studies of the stable analogues could provide great insight into the nature of the species formed in the kinetic folding process.

One approach to this is to study peptide fragments of the protein that are unable to fold to a native structure in the absence of the remainder of the polypeptide chain. The characterization of peptides has been widely used to search for conformational preferences in different regions of a sequence to provide clues as to the origin of early events in folding (Dyson *et al.* 1992; Kemmink & Creighton 1993; Waltho *et al.* 1993; Wu *et al.* 1993). A complementary strategy is to study fragments designed to probe interactions between regions of the protein in intermediate states (Oas & Kim 1988; Staley & Kim 1990). In the case of lysozyme we have synthesized the entire polypeptide chain in four segments with this purpose in mind. One result of particular interest is the discovery that the conformation of a peptide corresponding to the major β-sheet of the protein is able to adopt a variety of conformational states, depending on the solution conditions and its oligomeric state (Yang *et al.* 1994). This might provide clues as to the origin of misfolding events if indeed these are involved in the slow folding of the β-domain in the major pathway of folding. Detailed structural studies of the remaining three peptides by NMR are currently underway.

Another approach to the identification of species whose structures are potentially related to those of folding intermediates is to expose intact proteins to mild denaturing conditions. This strategy is of particular importance to our studies of the folding of lysozyme because α-lactalbumins, which are structural homologues of lysozymes, form partly folded states during equilibrium unfolding, for example at low pH (Kuwajima 1989). In studies using optical techniques such states (known as molten globules) have been correlated with early intermediates in the folding of both α-lactalbumins and lysozymes (Kuwajima 1989). Furthermore, NMR and ESI-MS studies have shown that these states are only very weakly protected from hydrogen exchange (the protection factors of the majority of amides are less than than ten (Baum *et al.* 1989; Chyan *et al.* 1993; Buck *et al.* 1994) and that the protected amides are located predominately in the α-domain (Baum *et al.* 1989; Chyan *et al.* 1993). In these respects they resemble most closely the molecules populated in the first few milliseconds of the lysozyme refolding experiments, suggesting that characterization of these stable states could provide considerable insight into the nature of very early events in the folding of

lysozyme. In this regard an important recent development is that the entire α-domain of human α-lactalbumin has been generated by protein engineering techniques and shown to form a stable state at neutral pH in the absence of denaturants with characteristics of the molten globule state of the intact protein (Peng & Kim 1994).

Structural studies using NMR of the molten globule state of α-lactalbumin have shown that despite clear evidence for secondary structure and some regions of local tertiary structure, the side-chains of at least the large majority of residues are substantially disordered (Baum *et al.* 1989; Alexandrescu *et al.* 1993). One interesting possibility is that this disorder, reflected in the broadening of resonances in the NMR spectrum (Baum *et al.* 1989; Alexandrescu *et al.* 1993), includes the slow interconversion of species as diverse as those thought to be responsible for the heterogeneity of the kinetic folding of lysozyme discussed above. Finally, recent work has indicated that states related to the α-lactalbumin molten globules may exist for some lysozymes. One of these, from the equine protein, is intriguing because this lysozyme is unusual in that it binds calcium; in this respect it has characteristics of both the lysozymes and α-lactalbumins (Van Dael *et al.* 1993). A second example has recently been found for human lysozyme (Haezebrouck *et al.* 1995) and it is interesting to speculate that this difference from the hen protein, whose equilibrium folding is highly cooperative under all the conditions explored so far, is related to the differences in the kinetic refolding behaviour of the two proteins.

Given that the molten globule states of α-lactalbumin appear to resemble the collapsed states formed during the kinetic refolding of lysozyme, the question of finding models for other states on the folding pathway arises. Detailed NMR studies of both the native and highly denatured states of lysozyme have been reported (Evans *et al.* 1991; Smith *et al.* 1993). The former correlates very well with the structure of the protein in the crystalline state (Blake *et al.* 1965; Smith *et al.* 1993). The structure of lysozyme in guanidinium chloride or urea appears to differ significantly from that of an archetypal random coil; of particular interest is the significant involvement of hydrophobic residues in residual structure (Broadhurst *et al.* 1991; Evans *et al.* 1991). This structure, however, does not appear to be persistent (Buck *et al.* 1994), nor to affect the events occurring during refolding (Kotik *et al.* 1995). Models for the more persistently structured states have proved harder to generate, although a partly folded state of lysozyme that has stable secondary structure in the absence of well-defined tertiary structure can be formed in TFE solution (Buck *et al.* 1993). In this state the secondary structure is most persistent in regions of the protein that are helical in the native state. This suggests that an environment which stabilizes structure in a polypeptide, such as might occur in the early stages of collapse, could generate native-like elements of secondary structure through a highly non-specific mechanism.

Although the structure of an analogue of the highly structured molten globules formed late in folding has not yet been characterized for lysozyme, structures of other proteins that are partly unfolded (at least locally) are now beginning to emerge which may share characteristics with such states (Feng *et al.* 1994; Redfield *et al.* 1994). At low pH, one such example is the four helix bundle protein, interleukin-4. Under these conditions, the protein has both near and far UV CD spectra that differ significantly from those of the native state, and an enhanced fluorescence in the presence of ANS (Redfield *et al.* 1994). It is, however, sufficiently native-like to allow its structure to be defined. This shows few significant differences from the fully native state, except the loss of several turns of one helix. Some 30 % of the polypeptide chain, however, is significantly disordered, as indicated most dramatically by nuclear relaxation measurements. Whether or not this has any direct relation to folding intermediates remains to be established, but it stresses the point that regions of secondary structure can be highly persistent, able to protect strongly against hydrogen exchange, whereas other regions of the native structure are still substantially unstructured.

5. RELATING *IN VIVO* AND *IN VITRO* FOLDING EVENTS

The folding process of a protein *in vitro* (see, for example, that of hen lysozyme described in figure 2) is likely to differ in many details from that in the cell. Apart from the obvious differences in the refolding conditions (temperature, pH, ionic strength and protein concentration), the presence of disulphide bonds at the onset of folding is artificial and is likely to affect significantly the stability of the various intermediate states, as it does the native state (Doig & Williams 1991; Cooper *et al.* 1992). A large part of this stabilization is likely to result from the reduction in the entropy of unfolded states relative to more highly folded ones, rather than from direct stabilization of specific structural features of particular states. Comparison of folding in the absence and presence of disulphide bonds, therefore, promises to throw considerable light on factors stabilizing intermediates and on the relation of the folding process *in vitro* to the situation occurring *in vivo*. A further complication in this comparison arises, however, as folding of proteins in the cell is assisted by proteins acting as catalysts such as protein disulphide isomerase and prolyl isomerase (Jaenicke 1993; Freedman *et al.* 1994). In addition, proteins such as the molecular chaperones, which are not thought to catalyse folding reactions but to increase the yield of the native protein by preventing aggregation of partly folded intermediates, are known to be important in folding in the cell (Ellis 1994; Hartl *et al.* 1994).

On the chaperone-assisted folding pathway the chaperonins (named GroEL in *Escherichia coli*) are thought to facilitate folding of their substrate proteins by providing a central cavity within which folding might take place in a protected environment (Langer *et al.* 1992; Chen *et al.* 1994). However, recent evidence using a mutant form of GroEL which is able to bind the substrate protein but not release its product, has

suggested that even in the chaperone-assisted pathway a substantial portion of the folding reaction might take place in the cytosol (Weissman *et al.* 1994). One of the major challenges, therefore, is to determine the relation between folding pathways determined *in vitro* with the situation occurring *in vivo*. Deciphering the molecular details of folding pathways *in vivo* is, however, no trivial task; not only because of the complexity of the GroEL tetradecamer (Braig *et al.* 1994) and the limited structural information about its interaction with the co-chaperone GroES (Landry & Gierasch 1994), but also in that elucidation of the structure of the bound substrate protein in the presence of the chaperonin is also required. From studies of ANS binding, intrinsic tryptophan fluorescence and proteinase sensitivity there is some evidence that the bound state has molten-globular characteristics (Martin & al 1991). Further characteristics of the nature of such a state, however, requires more detailed information such as that described above for intermediates formed *in vitro*.

How this can be achieved, however, is not obvious, as the size of the GroEL oligomer (> 800 kDa) is orders of magnitude larger than is amenable to direct NMR studies, precluding direct measurement of hydrogen exchange protection in the substrate protein by this usually powerful method.

One approach to overcoming this problem is to permit hydrogen exchange to take place within the intact complex, then to dissociate the components before recording the NMR spectrum of the released protein ligand. Such a procedure has been used to establish that cyclophilin bound to GroEL is substantially less stable than the native protein, and is in a state unable to protect amide hydrogens sufficiently to allow their detection (Zahn *et al.* 1994). However, this experiment could only put an upper limit of *ca.* 1000 on the protection factors in the bound state; such a value would not exclude any of the states on the lysozyme folding pathway except the native one. The success of the ESI-MS experiments described above to monitor hydrogen exchange has, therefore, prompted us to exploit this technique to characterize the structural details of a state of α-lactalbumin, which by reduction of a single disulphide bond and rearrangement of the remaining three disulphides has been shown to form a stable complex with GroEL (Hayer-Hartl *et al.* 1994; Robinson *et al.* 1994).

The idea behind this experiment is shown in figure 3. Hydrogen exchange is allowed to take place within the complex, but in this case the rate of hydrogen exchange is measured directly and in real time by introducing the intact complex into the mass spectrometer. Once this has occurred, the rapid loss of water from the protein in the gas phase prevents further exchange from taking place. By adjusting the characteristics of the mass spectrometer, conditions can be found under which the complex will dissociate in the mass spectrometer without perturbing the pattern of labelling in the protein ligand, allowing the mass of the α-lactalbumin molecules to be determined and hence the exchange process to be monitored. This experimental strategy has recently been put into practice (Robinson *et al.* 1994). The degree of protection measured for the

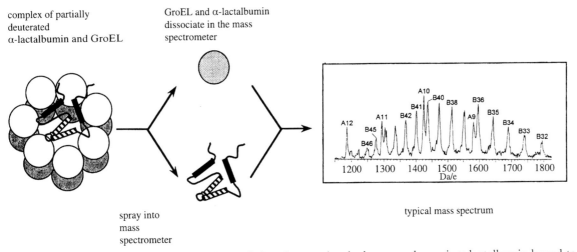

complex of partially
deuterated
α-lactalbumin and GroEL

GroEL and α-lactalbumin
dissociate in the mass
spectrometer

spray into
mass
spectrometer

typical mass spectrum

Figure 3. Schematic diagram of the experiment designed to monitor hydrogen exchange in α-lactalbumin bound to GroEL. The complex of GroEL with α-lactalbumin (which had previously been deuterated at all exchangeable sites) is diluted into H_2O and the rate of hydrogen exchange is measured directly by ESI-MS. In the diagram helices shaded with stripes denote regions protected from exchange, helices coloured black represent exchange labile regions. A typical mass spectrum of the complex is also shown. The charge state series labelled A arises from α-lactalbumin, that labelled B arises from the GroEL monomer (Robinson *et al.* 1994).

α-lactalbumin derivative bound to GroEL is much lower than that for the native state, in accord with the results for GroEL-bound cyclophilin (Zahn *et al.* 1994). What is even more striking, however, is that the bound protein is more highly protected than expected for a highly unfolded state. The extent of protection is very similar to that observed for a three-disulphide derivative of α-lactalbumin that exists in a molten globule state under the conditions of these experiments (Ewbank & Creighton 1993; Robinson *et al.* 1994). This suggests that GroEL interacts with folding intermediates having structures analogous to the disordered molten globules such as those formed in the early stages of the lysozyme folding pathway. In accord with this, recent experiments using intrinsic fluorescence measurements suggest that GroEL binds to the early collapsed states formed in the lysozyme folding pathway (M.P. Botwood & S.E. Radford, unpublished data).

Mass spectrometry has emerged, therefore, as a powerful method to complement NMR in monitoring hydrogen exchange. It provides novel information about the cooperativity of exchange processes, and about the populations of individual species having different numbers of protected amides. It can be applied to proteins of greater mass than can readily be studied by NMR, and requires only very small quantities of material. Information at the level of individual residues or regions of the sequence can be obtained by relating the results to those from complimentary NMR studies. In addition, we are exploring the possibility that specific hydrogen exchange information can also be obtained, at least in favourable cases, by using procedures that fragment the protein within the mass spectrometer. If such a procedure were to prove viable for protein molecules released from GroEL in the gas phase, then further details of the nature of the bound state are certain to emerge. Given that ESI-MS is in its infancy, the long term prospects for its widespread application in folding studies look excellent.

7. CONCLUSIONS

The studies of lysozyme and α-lactalbumin summarized here have revealed the power of using a battery of physical techniques to complement NMR in an attempt to describe the structural transitions occurring during the folding of a protein. The results have also enabled us to speculate on the nature of these states, and how this relates to the mechanism by which proteins are able to fold efficiently and accurately. Key ideas include the fact that a rapid collapse to a state with a rudimentary hydrophobic core could stabilize or even promote the formation of native-like secondary structure, and limit enormously the extent of conformational space that needs then to be explored to locate the native state itself. The observation of kinetically distinct populations of molecules emerging from the collapsed state supports the idea that only a proportion of molecules have attained a native-like topology in this process, and that the remainder fold more slowly because reorganizational events are needed before folding can proceed to the native state. The probability that misfolding occurs within a population of molecules is likely to be larger for longer polypeptide chains, and this might provide a considerable driving force towards the evolution of domains able to fold to a large extent independently of other regions of the structure. It would also be consistent with the fact that several small proteins, containing both β-sheets and α-helices, appear to fold extremely efficiently without the generation of molecules unable to fold rapidly from the collapsed state (Jackson & Fersht 1991; Briggs & Roder 1992; Kragelund *et al.* 1995).

As protein folding is usually fast, extremely accurate, and often rather insensitive to changes in conditions and even amino acid sequence, it seems likely that it must depend on rather simple universal characteristics of protein sequences and structure, such as their patterns of hydrophobic and hydrophilic residues (Dill *et al.* 1993). The importance of these patterns in determining characteristic folds is evident from their successful utilization in methods designed to predict which of a set of known structures a given sequence is most likely to favour (Bowie *et al.* 1991; Jones & Thornton 1993). Furthermore, it would be remarkable if the features of proteins that have developed to enable a protein to find a unique fold *in vivo* from an apparently overwhelming number of alternatives could by chance allow the same solution to be achieved by a fundamentally different route *in vitro*. In the case of α-lactalbumin discussed here, it is evident that GroEL binds to rather disordered but compact species corresponding most closely to species formed early during the folding of lysozyme. Indeed, it is tempting to speculate if this is the case that it might act to destabilize misfolded species and to avoid the trapping of such states that might otherwise fail to undergo further folding or might simply aggregate.

We acknowledge support from the EPSRC, BBSRC and MRC through the Oxford Centre for Molecular Sciences. S.E.R. is a Royal Society 1983 University Research Fellow. The research of C.M.D. is supported in part by an International Research Scholars Award from the Howard Hughes Medical Foundation and by a Leverhulme Fellowship from the Royal Society. We thank our collaborators who have contributed to this work in many ways, especially P.A. Evans (University of Cambridge) for his invaluable contribution to the *in vitro* studies of lysozyme and α-lactalbumin.

REFERENCES

Alexandrescu, A.T., Evans, P.A., Pitkeathly, M., Baum, J. & Dobson, C.M. 1993 Structure and dynamics of the acid-denatured molten globule state of α-lactalbumin: a two-dimensional NMR study. *Biochemistry* **32**, 1707–1718.

Baum, J., Dobson, C.M., Evans, P.A. & Hanley, C. 1989 Characterization of a partly folded protein by NMR methods: studies on the molten globule state of guinea pig α-lactalbumin. *Biochemistry* **28**, 7–13.

Blake, C.C.F., Koenig, D.F., Mair, G.A. & Sarma, R. 1965 Crystal structure of lysozyme by X-ray diffraction. *Nature, Lond.* **206**, 757–761.

Bowie, J.U., Lüthy, R. & Eisenberg, D. 1991 A method to identifying protein sequences that fold into a known three-dimensional structure. *Science, Wash* **253**, 164–170.

Braig, K., Otwinowski, Z., Hedge, R. *et al.* 1994 The crystal structure of the bacterial chaperonin GroEL at 2.8Å. *Nature, Lond.* **371**, 578–586.

Briggs, M.S. & Roder, H. 1992 Early hydrogen-bonding events in the folding reaction of ubiquitin. *Proc. natn. Acad. Sci. U.S.A.* **89**, 2017–2021.

Broadhurst, R.W., Dobson, C.M., Hore, P.J., Radford, S.E. & Rees, M.L. 1991 A photochemically induced dynamic nuclear polarization study of denatured states of lysozyme. *Biochemistry* **30**, 405–412.

Buck, M., Radford, S.E. & Dobson, C.M. 1993 A partially folded state of hen egg white lysozyme in trifluoroethanol: structural characterization and implications for protein folding. *Biochemistry* **32**, 669–678.

Buck, M., Radford, S.E. & Dobson, C.M. 1994 Amide hydrogen exchange in a highly denatured state of hen egg white lysozyme in urea. *J. molec. Biol.* **237**, 7–254.

Bychkova, V.E. & Ptitsyn, O.B. 1994 The molten globule in vitro and in vivo. *Chemtracts Biochem. molec. Biol.* **4**, 133–163.

Chaffotte, A., Guillou, Y. & Goldberg, M.E. 1992 Kinetic resolution of peptide bond and side chain far-UV circular dichroism during the folding of hen egg white lysozyme. *Biochemistry* **31**, 9694–9702.

Chen, S., Roseman, A.M., Hunter, A.S. *et al.* 1994 Location of a folding protein and shape changes in GroEL-GroES complexes imaged by cryo-electron microscopy. *Nature, Lond.* **371**, 261–264.

Chyan, C., Wormald, C., Dobson, C.M., Evans, P.A. & Baum, J. 1993 Structure and stability of the molten globule state of guinea pig α-lactalbumin: a hydrogen exchange study. *Biochemistry* **32**, 5681–5691.

Cooper, A., Eyles, S.J., Radford, S.E. & Dobson, C.M. 1992 Thermodynamic consequences of the removal of a disulphide bridge from hen lysozyme. *J. molec. Biol.* **225**, 939–943.

Darby, N.J., van Mierlo, C.P.M. & Creighton, T.E. 1991 The 5–55 disulphide intermediate in folding of BPTI. *FEBS Lett.* **279**, 61–64.

Denton, M.E., Rothwarf, D.M. & Scheraga, H.A. 1994 Kinetics of folding of guanidine-denatured hen egg white lysozyme and carboxymethyl (Cys6, Cys127)-lysozyme: A stopped flow absorbance and fluorescence study. *Biochemistry* **33**, 11225–11236.

Dill, K.A., Fiebig, K.M. & Chan, H.S. 1993 Cooperativity in protein folding kinetics. *Proc. natn. Acad. Sci. U.S.A.* **90** 1942–1946.

Dobson, C.M., Evans, P.A. & Radford, S.E. 1994 Understanding how proteins fold: the lysozyme story so far. *TIBS* **19**, 31–37.

Doig, A.J. & Williams, D.H. 1991 Is the hydrophobic effect stabilizing or destabilizing in proteins? The contribution of disulphide bonds to protein stability. *J. molec. Biol.* **217**, 389–398.

Dyson, H.J., Sayre, J.R., Merutka, G., Shin, H.-C., Lerner, R.A. & Wright, P.E. 1992 Folding of peptide fragments comprising the complete sequence of proteins. Models for initiation of protein folding. II. Plastocyanin. *J. molec. Biol.* **226**, 819–835.

Ellis, R.J. 1994 Roles of molecular chaperones in protein folding. *Curr. Opin. struct. Biol.* **4**, 117–122.

Evans, P.A. & Radford, S.E. 1994 Probing the structure of folding intermediates. *Curr. Opin. struct. Biol.* **4**, 100–106.

Evans, P.A., Topping, K.D., Woolfson, D.N. & Dobson, C.M. 1991 Hydrophobic clustering in non-native states of a protein: interpretation of chemical shifts in NMR spectra of denatured states of lysozyme. *Proteins Struct. Funct. Genet.* **9**, 248–266.

Ewbank, J.J. & Creighton, T.E. 1993 Structural characterization of the disulphide folding intermediates of bovine α-lactalbumin *Biochemistry* **32**, 3694–3707.

Eyles, S.J., Robinson, C.V., Radford, S.E. & Dobson, C.M. 1994 Kinetic consequences of removal of a disulphide bond on the folding of hen lysozyme. *Biochemistry.* **33**, 1534–1538.

Feng, Y., Sligar, S.G. & Wand, A.J. 1994 Solution structure of apocytochrome b562. *Nature struct. Biol.* **1**, 31–35.

Freedman, R.B., Hirst, T.R. & Tuite, M.F. 1994 Protein disulphide isomerase: building bridges in protein folding. *TIBS* **19**, 331–336.

Frydman, J., Nimmesgern, E., Ohtsuka, K. & Hartl, F.U. 1994 Folding of nascent polypeptide chains in a high molecular mass assembly with molecular chaperones. *Nature, Lond.* **370**, 111–117.

Gray, T.E. & Fersht, A.R. 1993 Refolding of barnase in the presence of GroE. *J. molec. Biol.* **232**, 1197–1207.

Haezebrouck, P., Joniau, M., Van Dael, H., Hooke, S.D., Woodruff, N.D. & Dobson, C.M. 1995 An equilibrium molten globule state of human lysozyme at low pH. *J. molec. Biol.* **246**, 384–389.

Hartl, F.-U., Hlodan, R. & Langer, T. 1994 Molecular chaperones in protein folding: the art of avoiding sticky situations. *TIBS* **19**, 20–25.

Hayer-Hartl, M., Ewbank, J.J., Creighton, T.E. & Hartl, F.U. 1994 Conformational specificity of the chaperonin GroE for the compact folding intermediates of α-lactalbumin. *EMBO J.* **13**, 3192–3202.

Hooke, S.D., Radford, S.E. & Dobson, C.M. 1994 The refolding of human lysozyme: a comparison with the structurally homologous hen lysozyme. *Biochemistry* **33**, 5867–5876.

Hughson, F.M., Wright, P.E. & Baldwin, R.L. 1990 Structural characterisation of a partly folded apomyoglobin intermediate. *Science, Wash.* **249**, 1544–1548.

Imoto, T., Johnson, L.N., North, A.C.T., Philips, D.C. & Rupley, J.A. 1972 *The enzymes* (ed. P.D. Boyer), 3rd edn, vol. 7, pp. 666–867. Orlando, Florida: Academic Press.

Itzhaki, L.S., Evans, P.A., Dobson, C.M. & Radford, S.E. 1994 Tertiary interactions in the folding pathway of hen lysozyme: kinetic studies using fluorescent probes. *Biochemistry* **33**, 5212–5220.

Jackson, G.S., Staniforth, R.A., Halsall, D.J. *et al.* 1993 Binding and hydrolysis of nucleotides in the chaperone catalytic cycle: implications for the mechanism of assisted protein folding. *Biochemistry* **32**, 2554–2563.

Jackson, S.E. & Fersht, A.R. 1991 Folding of chymotrypsin inhibitor 2. 1. Evidence for a two-state transition. *Biochemistry* **30**, 10428–10433.

Jaenicke, R. 1993 Role of accessory proteins in protein folding. *Curr. Opin. struct. Biology*, **3**, 104–112.

Jones, J. & Thornton, J. 1993 Protein fold recognition. *J. comp. molec. Des.* **7**, 439–456.

Kemmink, J. & Creighton, T.E. 1993 Local conformations of peptides representing the entire sequence of bovine pancreatic trypsin inhibitor and their roles in folding. *J. molec. Biol.* **234**, 861–878.

Koide, S., Dyson, H.J. & Wright, P.E. 1993 Characterization of a folding intermediate of apoplastocyanin trapped by proline isomerization. *Biochemistry* **32**, 12299–12310.

Kotik, M., Radford, S.E. & Dobson, C.M. 1995 The folding pathway of lysozyme denatured in guanidine hydrochloride and in DMSO. *Biochemistry* **34**, 1714–1724.

Kragelund, B.B., Robinson, C.V., Knudsen, J., Dobson, C.M. & Poulsen, F.M. 1995 A simple protein structural motif folds efficiently and cooperatively. *Biochemistry* (In the press.)

Kraulis, P.J. 1991 Molscript: a program to produce both detailed and schematic plots of protein structures. *J. appl. Cryst.* **24**, 946–950.

Kuwajima, K. 1989 The molten globule as a clue for understanding the folding and cooperativity of globular protein structures. *Proteins Struct. Funct. Genet.* **6**, 87–103.

Landry, S.J. & Gierasch, L.M. 1994 Polypeptide interactions with molecular chaperones and their relationship to in vivo protein folding. *A. Rev. Biophys. Biomol. Stuct.* **23**, 645–669.

Langer, T., Pfeifer, G., Martin, J., Baumeister, W. & Hartl, F.U. 1992 Chaperone-mediated protein folding: GroES binds to one end of the GroEL cylinder, which accommodates the protein substrate within its central cavity. *EMBO J.* **11**, 4757–4765.

Levinthal, C. 1968 Are there pathways for protein folding. *J. Chim. Phys.* **65**, 44–45.

Martin, J., Langer, T., Boteva, R., Schramel, A., Horwich, A.L. & Hartl, F.-U. 1991 Chaperonin-mediated protein folding at the surface of GroEL through a 'molten globule'-like intermediate. *Nature, Lond.* **352**, 36–42.

Miranker, A., Krupa, G.H., Robinson, C.V., Aplin, R.T. & Dobson, C.M. 1995 Site specific analysis of proteins by mass spectrometry. (Submitted.)

Miranker, A., Robinson, C.V., Radford, S.E., Aplin, R.T. & Dobson, C.M. 1993 Detection of transient protein folding populations by mass spectrometry. *Science, Wash.* **262**, 896–899.

Oas, T.G. & Kim, P.S. 1988 A peptide model of a protein folding intermediate. *Nature, Lond.* **336**, 42–47.

Peng, Z.Y. & Kim, P.S. 1994 A protein dissection study of a molten globule. *Biochemistry* **33**, 2136–2141.

Radford, S.E., Dobson, C.M. & Evans, P.A. 1992 The folding of hen lysozyme involves partially structured intermediates and multiple pathways. *Nature, Lond.* **358**, 302–307.

Redfield, C., Smith, R.A.G. & Dobson, C.M. 1994 Structural characterisation of a highly-ordered 'molten globule' at low pH. *Nature struct. Biol.* **1**, 23–29.

Robinson, C.V., Groß, M., Eyles, S.J. *et al.* 1994 Hydrogen exchange protection in GroEL-bound α-lactalbumin probed by mass spectrometry. *Nature, Lond.* **372**, 646–651.

Roder, H., Elöve, G.A. & Englander, S.W. 1988 Structural characterization of folding intermediates in cytochrome c by hydrogen exchange labelling and proton NMR. *Nature, Lond.* **335**, 700–704.

Sali, A., Shakhnovich, E. & Karplus, M. 1994 How does a protein fold? *Nature, Lond.* **369**, 248–251.

Senko, M.W. & McLafferty, F.W. 1994 Mass spectrometry of macromolecules: has its time now come? *A. Rev. Biophys. Biomolec. Struc.* **23**, 763–785.

Smith, L., Sutcliffe, M.J., Redfield, C. & Dobson, C.M. 1993 Structure of hen lysozyme in solution. *J. molec. Biol.* **229**, 930–944.

Staley, J.P. & Kim, P.S. 1990 Role of a subdomain in the folding of BPTI. *Nature, Lond.* **344**, 685–688.

Thirumalai, D. & Guo, Z. 1995 Nucleation mechanism for protein folding and theoretical predictions for hydrogen-exchange labelling experiments *Biopolymers* **35**, 137–140.

Udgaonkar, J.B. & Baldwin, R.L. 1988 NMR evidence for an early framework intermediate on the folding pathway of ribonuclease A. *Nature, Lond.* **335**, 694–699.

Van Dael, H., Haezebrouck, P., Morozova, L., Arico-Muendel, C. & Dobson, C.M. 1993 Partially folded states of equine lysozyme: Structural characterization and significance for protein folding. *Biochemistry* **32**, 11886–11894.

Waltho, J.P., Feher, V.A., Merutka, G., Dyson, H.J. & Wright, P.E. 1993 Peptide models of protein folding initiation sites. 1. Secondary structure formation by peptides corresponding to the G- and H-helices of myoglobin. *Biochemistry* **32**, 6337–6347.

Weissman, J.S., Kashi, Y., Fenton, W.A. & Horwich, A.L. 1994 GroEL-mediated protein folding proceeds by multiple rounds of binding and release of nonnative forms. *Cell* **78**, 693–702.

Wu, L.C., Laub, P.B., Elöve, G.A., Carey, J. & Roder, H. 1993 A non-covalent peptide complex as a model for an early folding intermediate of cytochrome c. *Biochemistry* **32**, 10271–10276.

Yang, J.J., Pitkeathly, M. & Radford, S.E. 1994 Far UV-circular dichroism reveals a conformational switch in a peptide fragment from the β-sheet of hen lysozyme. *Biochemistry* **33**, 7345–7353.

Zahn, R., Spitzfaden, C., Ottiger, M., Wüthrich, K. & Plückthun, A. 1994 *Nature, Lond.* **368**, 261–265.

Initial studies of the equilibrium folding pathway of staphylococcal nuclease

YI WANG, ANDREI T. ALEXANDRESCU AND DAVID SHORTLE

The Department of Biological Chemistry, The Johns Hopkins University School of Medicine, 725 North Wolfe Street, Baltimore, Maryland 21205, U.S.A.

SUMMARY

Spectroscopic methods were used to examine the sequential build up of structure in the denatured state of staphylococcal nuclease. The 'free energy distance' between the native and denatured states was manipulated by altering conditions in solution (for example altering urea or glycerol concentration) and by changing the amino acid sequences. Initial studies employed a fragment of nuclease, referred to as $\Delta131\Delta$, which lacks six structural residues from the amino terminus and one structural residue from the carboxy-terminus. Nuclear magnetic resonance analysis of this fragment in solution revealed a modest quantity of dynamic structure which is native-like in character. With the addition of urea, 12 new H_N peaks appeared in the ^1H–^{15}N correlation spectrum, presumably as a result of the breakdown of residual structure involving the first three beta strands. With the addition of glycerol, there was a rapid increase in the quantity of beta sheet structure detected by circular dichroism spectroscopy. At very high glycerol concentrations, an increase in helical structure became apparent. These data in addition to previously published results suggest that: (i) a beta-meander (strands $\beta1$-$\beta2$-$\beta3$) and the second alpha helix ($\alpha2$) are among the most stable local structures; (ii) the five-strand beta-barrel forms in a reaction which does not require the presence of several other native substructures; and (iii) the last step on the equilibrium folding pathway may be the formation and packing of the carboxy terminal alpha helix ($\alpha3$) to give the native state.

1. THEORETICAL FRAMEWORK

Contrasting the very rapid rate of spontaneous protein folding with the huge number of possible conformations that would have to be searched by a random folding mechanism, the need to invoke some sort of hierarchical mechanism for assembling native protein structure seems unavoidable (Levinthal 1968). Perhaps the simplest such mechanism would involve a series of clustering events that bring together successive segments of the protein chain as each segment assumes its correct, native-like structure. By proceeding through a series of intermediates which contain only native-like substructures, such a mechanism could efficiently funnel the chain into the native state, avoiding the necessity for breaking down or rearranging any non-native structures. Recent studies of denatured proteins, partly folded proteins and molten globules do in fact suggest that most of their structure is native-like (Matthews 1993; Shortle 1993). If proteins do fold by such a simple hierarchical mechanism, important goals for experimentalists will be to identify the various intermediate substructures formed as folding progresses towards the native state and to elucidate the chain–chain interactions which drive their formation and subsequent merger into higher order structures. Using the hypothetical folding diagram in figure 1, this goal can be restated in more concrete terms.

First, the individual steps in the clustering diagram, which involve combination of two substructures, must be defined and placed in hierarchical order. Second, the free energy change (ΔG) for each clustering step must be quantitated by measuring the relative concentrations of intermediates under specified sets of conditions. Finally, the physical chemical interactions that determine ΔG for each step must be defined.

Data of this type collected for several simple proteins would undoubtedly lead to many quantitative insights into the mechanisms by which amino acid sequence specifies protein structure. This framework for describing the folding of proteins focuses on the structural and free energy differences between intermediates, and not on the kinetics of their formation. As a result, time is not given the central role it plays in the kinetic approach to analysing the mechanism of protein folding. Instead of describing the pathway of structure formation as a function of time, such an equilibrium-based approach could be used to define an 'equilibrium folding pathway', one in which the appearance of structure is examined as a function of the free energy distance between a polypeptide chain blocked in its efforts to fold and the native state that it is attempting to reach. To vary this free energy distance, two distinctly different types of manipulation could be used to raise the free energy of the native state by controlled amounts relative to that of the denatured state.

Phil. Trans. R. Soc. Lond. B (1995) **348**, 27–34
Printed in Great Britain

27

© 1995 The Royal Society

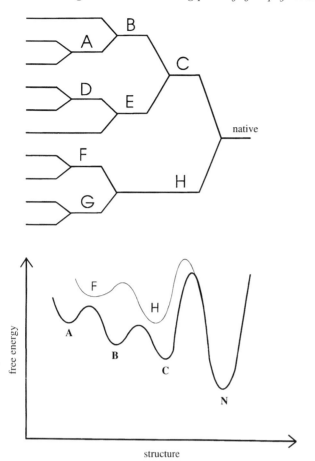

Figure 1. Top panel: hierarchical clustering diagram showing a series of hypothetical steps in the folding of a protein. A–H indicate folding intermediates, with increasing amounts of native-like structure proceeding from left to right. Bottom panel: free energy diagram of folding along two branches shown above involving several hypothetical intermediates.

First, disruption of native state interactions locally through changes in amino acid sequence increases the free energy of the native state with respect to the denatured state (Shortle 1992). Within the context of the model of structure assembly outlined in figure 1, a reasonable working assumption is that mutations will increase the free energy of only those partly folded intermediates that are stabilized by interactions which include the residue(s) at the site of the mutation (see figure 2a). In summary, formation of two different equilibrium intermediates will involve independent events unless they share a common substructure; intermediates that do not contain the interaction which has been disrupted by the mutation will not be affected. With this basis it should be possible to disrupt single branches on the hierarchical folding diagram, leaving other branches basically undisturbed, and by characterizing the structures of a large number of mutant denatured proteins (that have different sets of interactions disrupted) it may be possible to trace out the entire diagram. Alternatively, multiple native state interactions can be disrupted through changes in physical conditions of the protein solution, such as temperature, pH or solvent composition. When the free energy of the native state is increased by one of these variables, the effects on partly folded forms will be more universal, dependent on the physical chemistry

underlying that variable (see figure 2b). For example, when the pH is lowered below 3–5, the high concentration of protons raises the free energy of protein structures through increasing the entropy cost of maintaining unprotonated side chains and by increasing electrostatic repulsion involving positively charged groups (Tanford 1970). This will specifically disrupt substructures in proportion to how many unprotonated sidechains they bury and/or how many charged groups are brought into positions where they repel each other. Alternatively, when denaturants such as urea and guanidine hydrochloride are added, the free energy of protein structures is increased in proportion to the amount of nonpolar surface area that becomes solvent inaccessible on formation of that structure (Tanford 1970; Timasheff 1992). Instead of attempting to follow the very rapid appearance of structure by fast methods after the initiation of protein folding, experiments within this framework can employ slower, but higher resolution physical methods such as nuclear magnetic resonance (NMR) spectroscopy. Although this strategy for analysing protein folding is based on several assumptions, it provides a framework for interpreting structural data obtained under a series of equilibrium conditions. To evaluate the overall merit of this approach, our laboratory has begun NMR characterization of the structure that persists in staphylococcal nuclease after it has been denatured (i.e. in a state where it is unable to fold completely to the native state) by mutations or by physical agents. These data will be integrated into a consistent model – an equilibrium folding pathway – based on the expectations that: (i) the structure in a majority of intermediates will be native-like; and (ii) clustering steps are independent unless they involve intermediates that share a common substructure. In this report, previous NMR studies of a large, denatured fragment of nuclease are extended by monitoring the structural consequences of changing the free energy distance with urea and the stabilizing solute glycerol.

2. DATA

(a) *Δ131Δ: A low density denatured state of staphylococcal nuclease*

A fragment of staphylococcal nuclease, Δ131Δ, which is 131 residues in length but missing one structural residue from the carboxy terminus, and six structural residues from the amino terminus has been estimated to be denatured more than 99% of the time at 32 °C and pH 5.3 (Alexandrescu *et al.* 1994). Yet, upon addition of Ca^{2+} and substrate DNA (or inhibitory nucleotides), this protein can readily refold to the native state with essentially normal catalytic activity. Based on the backbone and side chain assignments of 105 of the 131 residues plus additional NMR data, a low resolution model of Δ131Δ has been proposed (Alexandrescu *et al.* 1994).

Figure 3 shows the principal structural features identified, which include: (i) a significant population of Type I and I′ reverse turns at 83–86 and 94–97, respectively; (ii) approximately 30% population of the

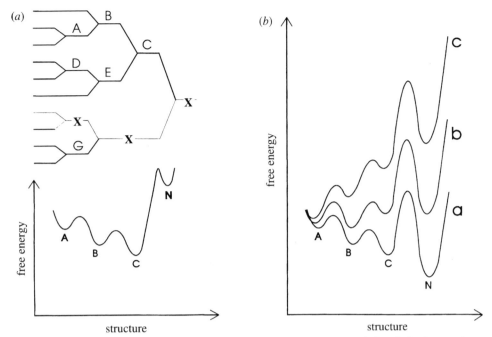

Figure 2. (*a*) Effects of mutating an amino acid residue that disrupts a chain–chain interaction present in intermediates F and H and in the native state. 'X' indicates that a clustering step in folding has been blocked by the mutation. Formation of intermediates A–E and G is unaffected by the mutation because these intermediates do not contain the disrupted interaction. (*b*) Effects on the free energy diagram of altering the solvent composition with denaturants (e.g. urea) and stabilizing solutes (e.g. glycerol). For denaturants/stabilizers, the free energy of a protein structure is raised/lowered, respectively, in proportion to the amount of solvent accessible surface area buried. Therefore the native state and intermediates with large amounts of structure will be destabilized/stabilized to a greater degree than intermediates with small amounts of structure by increasing the denaturant/stabilizer concentration in the order a–b–c/c–b–a, respectively.

second alpha helix; (iii) 10–20% population of alpha helix 1; and (iv) less than 10% population of alpha helix 3. Although no direct evidence of partial or transient beta strand formation has been found for any of the five strands that form the small Greek-key beta barrel in nuclease, 19 of the 24 H_N resonances corresponding to the first three beta strands are missing from the ^1H–^{15}N correlation spectrum (Alexandrescu *et al.* 1994). Because proton exchange rates under these solution conditions should be less than 1 s^{-1} for fully exposed amides (Molday *et al.* 1972) and as very strong resonances are observed for parts of the chain that have been shown to be solvent exposed (Alexandrescu & Shortle 1994), these resonances are unlikely to have been eliminated by proton exchange. Instead, the most probable explanation for severe broadening of a continuous segment of residues is conformational exchange on an intermediate timescale.

(*b*) *Δ131Δ: structure as a function of urea concentration*

To test the assumption that strands β1-β2-β3 are undergoing conformational exchange between an unfolded and a folded structure with very different values of chemical shift, the conditions of solution were made less favorable for structure formation by addition of a denaturant (see figure 2*b*). The ^1H–^{15}N spectrum of Δ131Δ was obtained at a series of urea concentrations from 0 M to 7 M in 1 M increments. Inspection of these spectra revealed several interesting features.

Those peaks present in the absence of urea exhibited only slight changes in their ^1H and ^{15}N chemical shifts between 0 M and 7 M urea. Whereas the large majority of peaks showed no significant change in intensity, 12 new peaks emerged, most of which first appeared at 3 or 4 M urea and increased in intensity at higher concentrations (see figure 4). One of these peaks falls in the range of H_N and ^{15}N chemical shifts that is diagnostic of a glycine residue in a random configuration. Because all glycines except Gly20 have been previously identified in Δ131Δ, this new peak can be assigned with confidence to Gly20, a residue which occupies the second position in a reverse turn connecting strands β1 and β2.

For five residues (Lys24, Leu25, Tyr27, Gly29 and Met32) in the peptide segment corresponding to β1-β2-β3, the H_N peaks have been assigned and in most cases these resonances are found to be significantly weaker than average. Unfortunately, except for Gly29, these peaks lie within crowded regions of the ^1H–^{15}N correlation spectrum and cannot be tracked with confidence between spectra collected at increasing urea concentrations to ascertain if they increase in intensity, as might be expected. In the case of Gly29, however, no significant increase in peak intensity was found, even at 7 M urea.

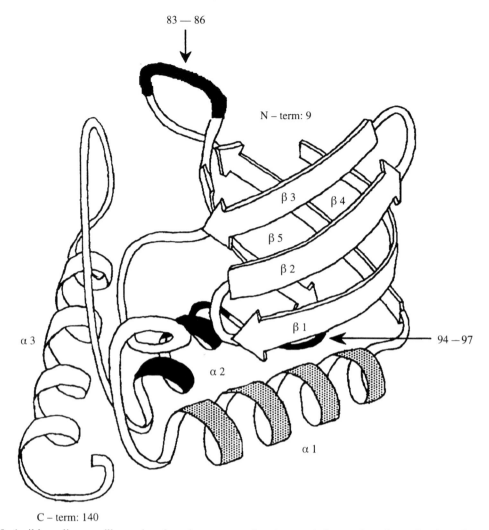

83 — 86

N – term: 9

β 3   β 4

β 5

β 2

β 1

94 — 97

α 3

α 2

α 1

C – term: 140

Figure 3. A ribbon diagram illustrating the substructures of native staphylococcal nuclease that have been detected in the large nuclease fragment Δ131Δ. The shaded regions indicate the elements of native-like residual structure (black, most persistent; grey, less persistent) based on secondary chemical shifts, NOEs and $^3J_{N_\alpha}$ coupling constants. Few of the residues in the first three beta strands β1-β2-β3 have been assigned. (From Alexandrescu *et al.* 1994).

(c) Δ131Δ: structure as a function of glycerol concentration

To study the response of Δ131Δ to conditions which are more favourable for folded structure (see figure 2b), the stabilizing solute glycerol was added to solutions of Δ131Δ at 32 °C and pH 5.3. As can be seen in figure 5a, large changes in the circular dichroism spectra suggestive of major increases in secondary structure occur between 0% and 70% glycerol (by volume). The fraction of molecules that have refolded to the native state can be monitored by the intrinsic fluorescence of tryptophan 140 (Shortle & Meeker 1986) with the result that less than 5% of molecules are in the native state at 30% glycerol, whereas approximately 15% have refolded in 70% glycerol (data not shown).

Further evidence that the additional structure induced at low concentrations of glycerol is not a consequence of complete folding to the native state is shown in figure 5b. These circular dichroism (CD) spectra of Δ131Δ were collected with increasing concentrations of the competitive inhibitor 3,5 di-phosphothymidine (pdTp) in the presence of a fixed concentration of Ca²⁺. The nucleotide plus one calcium ion bind to the active site of nuclease with very high affinity (Serpersu *et al.* 1986) and refold Δ131Δ by substrate stabilization of the native state. As can be seen, these spectra, which represent contributions from varying fractions of denatured and native states, look distinctly different from the spectra obtained in glycerol. These observations can be interpreted more quantitatively by calculating the difference spectra between Δ131Δ in the presence and absence of these two types of structure stabilizers. In figure 5c, the difference spectrum for 30% glycerol minus 0% glycerol displays the characteristic minimum at 215 nm typically seen for pure beta sheet structures (Johnson 1988), whereas the difference spectrum for 70% glycerol minus 30% glycerol has the minima at 222 nm and 208 nm diagnostic of alpha helix. In contrast, the difference spectra obtained as a function of pdTp are distinctly different from these two, exhibiting a strong minimum at 222 nm without a pronounced minimum at 208 nm. This suggests that alpha-helix and beta sheet are forming simultaneously rather than consecutively, as expected for one-step refolding to the native state. The 30%–0% difference spectrum sug-

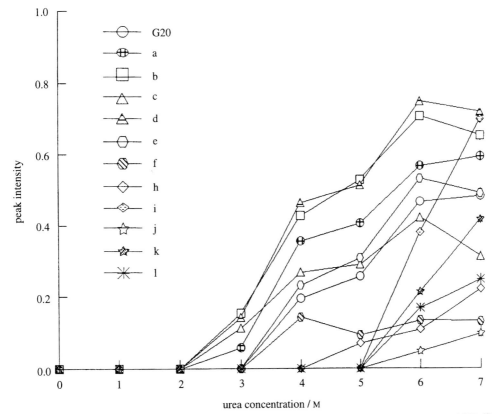

Figure 4. The relative intensity of twelve 'new' H_N resonances of $\Delta131\Delta$ resolved in a series of ^1H–^{15}N–HSQC (Messerle *et al.* 1989) spectra collected at different concentrations of urea at pH 5.3 and 32 °C. Relative intensity is the measured peak height normalized to the height of the H_N of Serine 128. Peaks that were not present in 0 M urea and that were clearly resolved from all other peaks are shown. Only one of these, Glycine 20, has been assigned.

gests the transition at lower glycerol concentrations predominantly involves the formation of beta sheet. When the assigned H_N resonances are followed in the ^1H–^{15}N correlation spectrum as a function of glycerol, all but one (Ile92) of the peaks retain essentially the same position in the spectrum from 0% to 30% glycerol. However, all peaks undergo significant reductions in intensity, albeit at very different rates.

As seen in figure 6, the H_N resonances from beta strands β4 and β5 diminish in intensity most rapidly with increasing glycerol. The H_N resonances from the third alpha helix α3 and the catalytic loop (residues 43–54; data not shown) decrease by a relatively small amount. Finally, the H_N resonances from the first and second alpha helix α1 and α2 display a behaviour which is intermediate between these two extremes. It is expected that concentrations of glycerol above 5–10% significantly increase the solution viscosity, which will increase the rotational correlation time and thereby reduce T_2 for both H_N and ^{15}N directly via effects on dipolar–dipolar and chemical shift anisotropy relaxation mechanisms (Abragam 1961; Wagner 1993).

However, when such direct broadening effects are measured on wild-type nuclease, they are found to be relatively small. For example, quantitation of the peak intensities of 23 well resolved H_N peaks of native nuclease (molecular mass 17000) in the presence and absence of 30% glycerol (at 32 °C, pH 5.3) revealed an average reduction of peak intensity to 0.64 of the initial value (data not shown). Because the apparent rotational correlation times of residues in $\Delta131\Delta$ have been

demonstrated to be significantly shorter than that of wild-type nuclease (Alexandrescu & Shortle 1994), a 36% decrease in peak intensity would appear to represent an upper estimate of the direct effect of glycerol on T_2 and the observed line widths.

As mentioned above, only slight changes in ^1H and ^{15}N chemical shifts were observed for the H_N resonances in $\Delta131\Delta$ (except Ile92) as the glycerol was increased. From this observation alone, it can be concluded that: (i) conformational exchange is unlikely to be fast on the chemical shift time scale; and (ii) the observed chemical shifts at all glycerol concentrations are those of an unstructured state which is more highly populated than the state with which it exchanges. Given the extensive broadening of so many residues, the H_N chemical shifts in the less populated, more structured state state must be significantly different for essentially all residues in $\Delta131\Delta$.

3. CONCLUSIONS

The most interesting observation reported here is that, upon addition of the stabilizing solute glycerol, the polypeptide chain of $\Delta131\Delta$ undergoes a major transition to a structure with less alpha helical content than the native state. A prominent feature of this transition appears to be very extensive exchange broadening of the H_N resonances, suggesting that the final structure is native-like. This conclusion is based on the fact that the chemical shift environments of many resonances are distinctly different from their

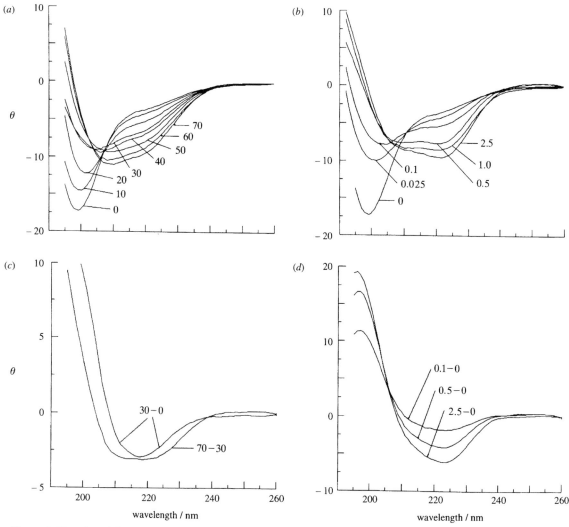

Figure 5. Far ultraviolet circular dichroism spectra of Δ131Δ. The x axis is wavelength of ultraviolet light in nm, the y axis is the mean residue molar ellipticity ($\times 10^{-3}$) (θ). (a) As a function of glycerol concentration, each curve is labelled with the % glycerol. (b) As a function of 3,5 diphosphothymidine concentration (pdTp) at pH 6.5 and 5 mM CaCl$_2$, after subtraction of a blank spectrum collected with the same concentrations of pdTp and Ca^{2+} but without protein, each curve is labelled with the pdTp concentration in mM. (c) Calculated difference spectra from part (a), 30–0 is the difference between 30% and 0% glycerol; 70–30 is the difference between 70% and 30% glycerol. (d) Calculated difference spectra from panel (b) 0.1–0 (0.5–0, 2.5–0) represent the difference between 0.1 (0.5, 2.5) mM pdTp and 0 mM pdTp.

random, highly averaged values. Because this transition is rapidly promoted by increasing concentration of glycerol and because glycerol affects conformational transitions in proportion to the amount of nonpolar surface area that undergoes burial (Gekko *et al.* 1981), the structure formed in this transition must involve the burial of a large number of hydrophobic residues. At concentrations of glycerol less than 35%, this state lacks much of the alpha helical structure found in the native state. As the glycerol concentration is raised even further, there is a specific increase in alpha helical structure. From these observations, the simplest conclusion is that the five-strand beta barrel structure of nuclease forms initially in the absence of some of the alpha helices. At higher glycerol concentrations, these helices become increasingly populated and pack against the beta barrel to give the native or a near-native structure. Previous NMR analysis of Δ131Δ in the absence of glycerol has revealed no evidence for beta structure formation by chain segments corresponding

to beta strands β4 and β5 (Alexandrescu *et al.* 1994). However, the H$_N$ resonances of all of the residues in β1 and over half of the residues in β2 and β3 are missing from the ^{1}H–^{15}N correlation spectrum. In view of the relatively slow (approximately 1 s^{-1}) proton exchange rate predicted at the temperature and pH used, the most likely explanation for this extensive broadening is intermediate exchange between an unstructured state and a structured state, with the beta meander formed by β1-β2-β3 as the obvious candidate for the structured state. Consistent with such a broadening mechanism is the appearance of 12 new peaks upon adding the denaturant urea to concentrations of 3 M and above. One of these peaks can be assigned to Gly20, a residue which plays a central role in this beta meander. Broadening of the H$_N$ resonances corresponding to the third alpha helix α3 only becomes signficant at high glycerol concentrations. Because the decrease in peak intensity of these residues is approximately the same as is observed for native WT nuclease

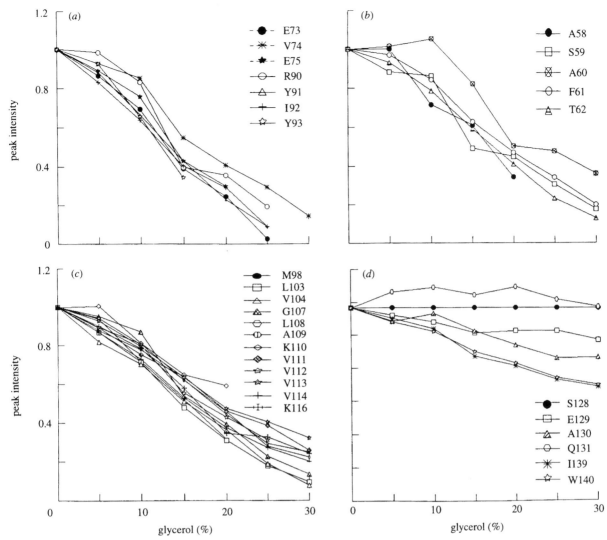

Figure 6. The relative intensity of the H_N resonances of $\Delta 131\Delta$ resolved in a series of ^1H–^{15}N–HSQC spectra (Messerle *et al.* 1989) collected a different glycerol concentrations at pH 5.3 and 32 °C. Relative intensity is the measured peak height normalized to the height of the H_N of Serine 128. (*a*) Residues corresponding to beta strands β4–β5. (*b*) Residues corresponding to helix α1. (*c*) Residues corresponding to helix α2 plus an extended segment. (*d*) Residues corresponding to helix α3.

in glycerol, the observed broadening could be a consequence of increasing solution viscosity rather than formation of an alpha helix. Previous ^{15}N relaxation studies clearly indicated the residues that comprise this chain segment are among the most mobile and unconstrained in $\Delta 131\Delta$ (Alexandrescu & Shortle 1994). The disinclination of this structure to form was also noted in a structural analysis of a quite different denatured form of staphylococcal nuclease, involving a large fragment missing five structural residues from the carboxy terminus and none from the amino terminus (Shortle & Abeygunawardana 1993). In this case, the five strand beta barrel was demonstrated to be intact, yet the α3 segment displayed extremely sharp backbone resonances with random-coil chemical shifts, suggesting a very high mobility. Thus, several lines of evidence place the packing of helix α3 against the remainder of the molecule as one of the last steps, if not the final step, in the assembly of the wild-type native structure. The results presented here, coupled with previous solution studies of $\Delta 131\Delta$ (Alexandrescu *et al.* 1994; Alexandrescu & Shortle

1994) allow a tentative, partial equilibrium folding pathway to be defined for staphylococcal nuclease, as shown below.

Future studies of additional mutant forms of staphylococcal nuclease of normal length should allow a more complete dissection of this hierarchy of clustering steps plus an estimation of the various values of ΔG involved.

We thank Joel Gillespie for help with the CD spectroscopy. This work was supported by NIH Grant GM34171 to D.S.

REFERENCES

Abragam. A. 1961 *The principles of nuclear magnetism.* London: Oxford University Press.

Alexandrescu, A.T. & Shortle, D. 1994 Backbone dynamics

of a highly disordered 131 residue fragment of staphylococcal nuclease. *J. molec. Biol.* **242**, 527–546.

Alexandrescu, A.T., Abeygunawardana, C. & Shortle. D. 1994 Structure and dynamics of a denatured 131 residue fragment of staphylococcal nuclease: a heteronuclear NMR study. *Biochemistry* **33**, 1063–1072.

Gekko, K. & Timasheff, S.N. 1981 Mechanism of protein stabilization by glycerol: preferential hydration in glycerol-water mixtures. *Biochemistry* **20**, 4667–4676.

Johnson, W.C. Jr 1988 Secondary structure of proteins through circular dichroism spectroscopy. *A. Rev. biophys. Biophys. Chem.* **17**, 145–166.

Levinthal, C. 1968 Are there pathways for protein folding? *J. chim. Phys.* **65**, 44–45.

Matthews, C.R. 1993 Pathways of protein folding. *A. Rev. Biochem.* **62**, 653–683.

Messerle, B.A., Wider, G., Otting, G., Weber, C. & Wuthrich, K. 1989 Solvent suppression using a spin-lock in 2D and 3D NMR spectroscopy with H_2O solutions. *J. magn. Reson.* **85**, 608–613.

Molday, R.S., Englander, S.W. & Kallen, R.G. 1972 Primary structure effects on peptide group hydrogen exchange. *Biochemistry* **11**, 150–159.

Serpersu, E.H., Shortle, D. & Mildvan, A.S. 1986 Kinetic and magnetic resonance studies of effects of genetic substitution of a Ca^{2+}-liganding amino acid in staphylococcal nuclease. *Biochemistry* **25**, 68–77.

Shortle, D. 1992 Mutational studies of protein structures and their stabilities. *Q. Rev. Biophys.* **25**, 205–250.

Shortle, D. 1993 Denatured states of proteins and their roles in folding and stability. *Curr. Opin. struct. Biol.* **3**, 66–74.

Shortle, D. & Meeker, A.K. 1986 Mutant forms of staphylococcal nuclease with altered patterns of guanidine hydrochloride and urea denaturation. *Proteins: Struct. Funct. Genet.* **1**, 81–89.

Shortle, D. & Abeygunawardana, C. 1993 NMR analysis of the residual structure in the denatured state of an unusual mutant of staphylococcal nuclease. *Structure* **1**, 121–134.

Tanford, C. 1970 Protein denaturation. Part C. Theoretical models for the mechanism of denaturation. *Adv. Protein Chem.* **24**, 1–95.

Timasheff, S.N. 1992 Water as ligand: preferential binding and exclusion of denaturants in protein unfolding. *Biochemistry* **31**, 9857–9864.

Wagner, G. 1993 NMR relaxation and protein mobility. *Curr. Opin. struct. Biol.* **3**, 748–754.

Kinetic and equilibrium folding intermediates

O. B. PTITSYN[1,2], V. E. BYCHKOVA[1] AND V. N. UVERSKY[1]

[1] *Institute of Protein Research, Russian Academy of Sciences, Pushchino, Moscow Region, 142292, Russia*
[2] *Laboratory of Mathematical Biology, National Cancer Institute, NIH, Bethesda, Maryland 20892, U.S.A.*

SUMMARY

Our recent experiments on the molten globule state and other protein folding intermediates lead to following conclusions: (i) the molten globule is separated by intramolecular first-order phase transitions from the native and unfolded states and therefore is a specific thermodynamic state of protein molecules; (ii) the novel equilibrium folding intermediate (the 'pre-molten globule' state) exists which can be similar to the 'burst' kinetic intermediate of protein folding; (iii) proteins denature and release their non-polar ligands at moderately low pH and moderately low dielectric constant, i.e. under conditions which may be related to those near membranes.

1. INTRODUCTION

The main difficulty of protein folding is to avoid 'traps' (local energy minima) in which protein can fall down. The 'framework' model of protein folding (Ptitsyn 1973, 1987) assumes that protein avoids this difficulty by folding step-by-step in such a way that the results of the each step are not reconsidered but just fastened at subsequent steps. The framework model has predicted two main kinetic intermediates of protein folding. In the first intermediate a secondary structure fluctuating around its native position is already formed, whereas the second intermediate has much more stable secondary structure and the main features of the native tertiary fold (i.e. the crude mutual positions of α-helices and β-strands).

Both these kinetic intermediates have been observed experimentally and described in detail. First, it was shown that secondary structure (Robson & Pain 1976) and a compact state (Creighton 1980) are formed before the tertiary structure. Secondly, in 1987 it was shown that protein folds through at least two kinetic intermediates. The first intermediate is formed in less than 10 ms and has a substantial fluctuated secondary structure (Gilmanshin & Ptitsyn 1987; Kuwajima *et al.* 1987; Elöve *et al.* 1992; Radford *et al.* 1992). The second intermediate is formed within 0.1–1 s and has a globular shape (Semisotnov *et al.* 1987, 1992) and much more stable secondary structure (Elöve *et al.* 1992; Radford *et al.* 1992). By using the pulsed hydrogen exchange method (Baldwin 1993; Roder & Elöve 1994) and site-specific mutagenesis (Fersht 1993) it has been shown that the second kinetic intermediate (usually just preceding the rate-limiting step of folding) has many features of the three-dimensional structure of the native protein.

It was shown (Dolgikh *et al.* 1984; Baldwin 1993; Jennings & Wright 1993) that the second kinetic intermediate shares many features with an equilibrium intermediate, the molten globule state (Dolgikh *et al.* 1981), which is a typical state of protein molecules under mild denaturing conditions (for review, see

Ptitsyn 1992). It was shown also that the molten globule state is involved into some important physiological processes (for review, see Bychkova & Ptitsyn 1993).

This paper summarizes our recent results on the molten globule state and other equilibrium folding intermediates.

2. THE MOLTEN GLOBULE IS A SPECIFIC THERMODYNAMIC STATE OF PROTEIN MOLECULES

One of the most important questions about the molten globule state is whether it is a specific thermodynamic state of protein molecules or whether it is similar to either a slightly disordered native protein or a 'squeezed coil'. The criterion of a specific thermodynamic state is the presence of phase transition between that state and other ones. The most important are first-order phase transitions ('all-or-none' transitions) which are coupled with drastic change of at least one of the first derivatives of free energy (like enthalpy, number of 'absorbed' solvent molecules, etc.). It was shown that protein denaturation (i.e. loss of its activity accompanied by the loss of its rigid tertiary structure) is of all-or-none character, presenting the first example of intramolecular first-order phase transition (Privalov 1979). However, it was assumed (Privalov 1979, 1992) and widely accepted that all denatured states of proteins are identical or similar from the thermodynamical point of view.

To determine whether the molten globule is separated by all-or-none transitions from other states, we applied the well-known statistical physics concept that the slope of all-or-none transition in a small system is proportional to a number of units in that system (Hill 1968). Applied to this investigation, it means that the slope of an intramolecular transition must be proportional to molecular mass of a macromolecule.

Figure 1 shows experimental data on the slopes of urea- and guanidinum chloride (GdmCl)-induced N ⇔ MG and MG ⇔ U transitions (N, native state; MG,

Phil. Trans. R. Soc. Lond. B (1995) **348**, 35–41

Printed in Great Britain

35

© 1995 The Royal Society

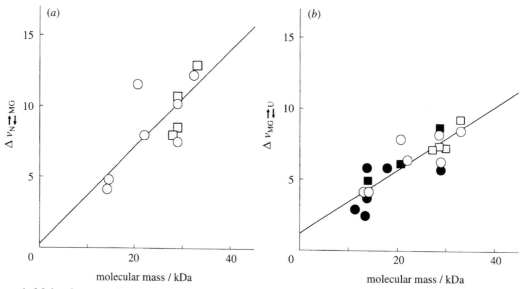

Figure 1. Molecular mass dependence of slopes for urea-induced (squares) and GdmCl-induced (circles) N⇔MG (a) and MG⇔U (b) transitions. Open symbols refer to three-state unfolding of native proteins (N⇔MG⇔U), while filled symbols refer to two-state unfolding of pH-induced MG state (MG⇔U). Slopes are presented as $\Delta\nu = \partial \ln K / \partial \ln a$ in the middle of transition, where K is equilibrium constant between two states of a protein molecule and a is activity of urea or GdmCl. Adapted from Ptitsyn & Uversky (1994).

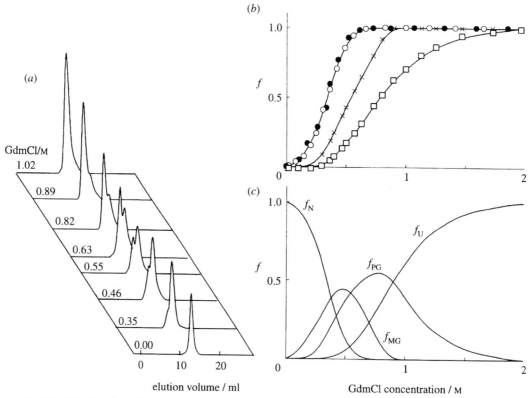

Figure 2. GdmCl-induced unfolding of *Staphylococcal β*-lactamase. (a) Elution profiles at different concentrations of GdmCl. (b) GdmCl dependence of activity (open circles) and molar elipticity at 270 nm (filled circles), as well as fraction of 'less compact' molecules (x) and their elution volume (open squares) in relative units. (c) GdmCl dependence of fractions of native (N), molten globule (MG), pre-molte globule (PG) and unfolded (U) protein molecules. Adapted from Uversky & Ptitsyn (1994).

molten globule state; U, unfolded state) in all small (one-domain) proteins studied up-to-date (Ptitsyn & Uversky 1994). The figure demonstrates that in both cases the slopes are proportional to the molecular masses of proteins. Thus, both N⇔MG and MG⇔U transitions are of all-or-none character, i.e. are intramolecular first-order phase transitions. The all-or-

none character of N⇔MG transition was indeed demonstrated even earlier for a temperature denaturation of α-lactalbumins which are in the MG state at high temperatures (Dolgikh *et al.* 1981, 1985).

The all-or-none character of MG⇔U transition has been demonstrated by the most direct method: bimodal distribution of protein elution volumes upon

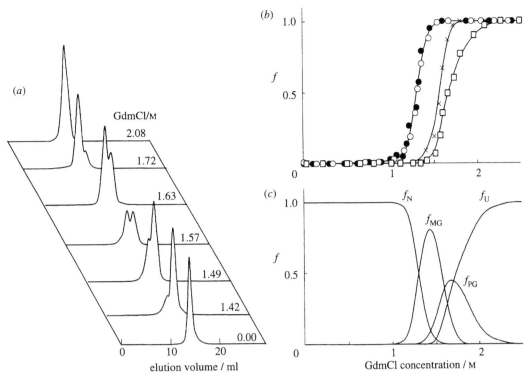

Figure 3. GdmCl-induced unfolding of bovine carbonic anhydrase B. All notations are the same as in figure 2. Adapted from Ptitsyn & Uversky (1995).

GdmCl-induced protein unfolding (Uversky *et al.* 1992; Uversky & Ptitsyn 1994, 1995). Figures 2*a* and 3*a* show this bimodal distribution for two studied proteins and present a clear evidence for an all-or-none transition. This transition occurs at concentrations of GdmCl substantially higher then the concentrations which lead to denaturation of these proteins (see figures 2*b* and 3*b*). Therefore it is not connected with protein denaturation but rather is the transition between two denatured states: a compact and a less compact. The first denatured state is non-active, has no rigid tertiary structure, but is compact, has a large content of secondary structure and strongly binds 1-anilino-naphtalene-8-sulphonate (ANS), thus meeting all usual criteria of the MG state. As to the second (less compact) denatured state, it is a mixture of at least two states with quite different volumes (see below).

The possible explanation of two equilibrium phase transitions in protein molecules is the existence of two levels of their three-dimensional structure: a large-scale order ('tertiary fold') which can exist already in the MG state (Peng & Kim 1994), and a short-scale order (rigid tertiary structure) which exists only in the native state. It is quite possible that two phase transitions in proteins correspond to the formation (or destruction) of these two levels of protein structures (Ptitsyn 1994).

3. EQUILIBRIUM 'PRE-MOLTEN GLOBULE' STATE: A POSSIBLE ANALOGUE OF THE FIRST KINETIC INTERMEDIATE

Figures 2 and 3 demonstrate an additional important feature of the equilibrium folding or unfolding of proteins. They show that an elution volume of less compact molecules substantially decreases with the increase in GdmCl concentration, which corresponds to a further increase of a hydrodynamic volume (cf. Palleros *et al.* 1993). This increase is much larger than the normal swelling of unfolded proteins in good solvents (Uversky & Ptitsyn 1994) and continues even at those GdmCl concentrations at which all compact (MG) molecules disappear. It can be explained only by the existence of two different 'less compact' states which are in a fast equilibrium: the partly folded 'pre-molten globule' (PG) state and the really unfolded state (Uversky & Ptitsyn 1994, 1995). For both studied proteins a hydrodynamic volume (V) of the pre-molten globule state $V_{PG} \lesssim 2.4\ V_N$ (and $\lesssim 1.5\ V_{MG}$), while $V_U \simeq 12V_N$. Far ultraviolet circular dichroism (UV CD) spectra show that the pre-molten globule state has a substantial secondary structure, i.e. the secondary structure of the MG state decreases in two steps: upon MG → PG and PG → U transitions. The pre-molten globule state also binds ANS although less strongly than the MG state. Thus, this new state belongs to the family of relatively compact and partly structured states of protein molecules. We have suggested that the pre-molten globule state may include an 'embryo' of the native-like tertiary fold (Uversky & Ptitsyn 1995).

Properties of this novel equilibrium state are similar to those of the first kinetic intermediate. In fact, this kinetic intermediate also is partly condensed (Kawata & Humaguchi 1991; Elöve *et al.* 1992), has a substantial amount of secondary structure (Kuwajima *et al.* 1993) and binds ANS (Semisotnov *et al.* 1991).

It is known that both the first kinetic intermediate and the kinetic MG state accumulate upon protein folding, i.e. are separated by high potential barriers both from each other and from the native state. The existence of two equilibrium phase transitions in proteins and the accumulation of two kinetic intermediates upon protein folding therefore perhaps have a common purpose: the independent formation of two levels of three-dimensional structure (Ptitsyn 1994).

4. FOLDING INTERMEDIATES UPON PHYSIOLOGICAL CONDITIONS

Physical studies of the molten globule state suggest that the molten globule preserves the main elements of native secondary structure and their crude mutual positions in three-dimensional space, but differs from the native state by a less tight packing of side chains (Shakhnovich & Finkelstein 1989), and by partial unfolding of loops and ends of a chain (Ptitsyn 1992). This state is almost ideal for a protein which have to adapt itself to different external conditions (like the conditions in a living cell), maintaining a memory on its overall architecture. It was the basis of our prediction that the molten globule state can exist in a living cell and can be involved in a number of physiological processes (Bychkova et al. 1988). Part of these predictions – especially recognition of the MG state by chaperones (Martin et al. 1991) and the role of the MG state in insertion of proteins into membranes (van der Goot et al. 1991) – has been confirmed experimentally. More recently a number of other physiological applications of the MG state have been shown or suggested (for a review, see Bychkova & Ptitsyn 1993).

In many cases folding intermediates in a cell are the kinetic intermediates trapped by chaperones just after protein biosynthesis before proteins can completely fold. It is known that chaperones transfer proteins in these intermediates states to corresponding compartments of a cell where they are released from chaperones, are translocated across membranes into the compartment and then acquire their completely folded structures (Schatz 1993; Hendrick & Hartl 1993). Another possible case may be proteins with mutations which prevent them from complete folding (see below).

However, there are also proteins, or their domains, which normally exist in the rigid ('native') state but must denature to fulfil their function. Examples include pore-forming domains of some toxins (van der Goot et al. 1992), as well as proteins that act as carriers for large non-polar ligands (see below). How these proteins can denature at physiological conditions (at usually neutral bulk pH, large ionic strength and normal temperatures) is not clear. However, we have to remember that a cell is transpired by membranes and cytoskeleton and contains a lot of proteins, nucleoprotein complexes and organelles. Thus, it can be compared to a thick Russian soup rather than to salted water which often is believed to be a proper model for a natural environment of proteins.

There are at least two factors which can lead to a protein denaturation near membrane surfaces. The first is that negative charges of membrane attract protons which leads to a local decrease of pH (van der Goot et al. 1991). The second is that an organic moiety of membranes decreases an effective value of local dielectric constant (Bychkova & Ptitsyn 1993).

A very crude model of a 'concert' action of these two effects may be a protein denaturation in water–organic mixtures at moderately low pH, although the states of denatured proteins in these cases may be different from their 'normal' folding intermediates (see below).

5. PROTEIN DENATURATION AND RELEASE OF NON-POLAR LIGANDS

A good example of this situation are protein carriers of large non-polar ligands. These ligands often are deeply buried into a rigid protein and tightly packed with non-polar groups of its core as it is the case, for instance, for retinol-binding protein (RBP). The release of these ligands may become possible only in the molten globule or other denatured state where ligands are less tightly packed with protein groups (Bychkova & Ptitsyn 1993).

It was shown (Bychkova et al. 1992) that retinol can be released from RBP at low pH (in the interval from 5.5 to 3.0 with the middle point about 4.5) and that the release of retinol is coupled with the protein transition into the MG state. This was the first evidence that the MG state can be involved in a target release of non-polar ligands. Of course, in water solutions both transitions occur at pH which are much lower than their physiological values. However, recently we have shown (V.E. Bychkova, A. Fantuzzi, A.E. Dujsekina, G.-L. Rossi & O.B. Ptitsyn, unpublished data) that both release of retinol and denaturation of RBP can be achieved at substantially higher pH in water–methanol mixtures. For example, the increase of methanol content from 0 to 30 and 50% shifts the ends of both these transitions from pH 3.0 to 4.5 and 7.0, respectively. As a result, a native RBP-retinol complex is stable only at pH > 5.5 and at methanol content less than 35%, which corresponds to average dielectric constants larger than 60. These experiments clearly demonstrate the importance of a 'concert' action of pH and organic environment on a function of protein coupled with its denaturation.

Another interesting example of a protein carrying large non-polar ligands is α-fetoprotein, which transfers unsaturated fatty acids and estrogens to embryonal and some cancer cells (for a review, see Abelev 1993). It was shown (Uversky et al. 1995) that α-fetoprotein also can be transformed into MG state at pH 3 in water solutions. In fact, figure 4 shows that this protein at pH 3.1 has a native-like far UV CD spectrum and native-like spectrum of Trp fluorescence, but has no co-operative temperature melting and strongly binds ANS, i.e. meets all usual requirements of the MG state. It is quite possible that the release of unsaturated fatty acids and other ligands from α-fetoprotein also is connected with the transformation of rigid 'native' molecules into molten globule or other denatured state.

Of course, proteins denatured in water–alcohol mixtures can be quite different from the folding intermediates in water. In fact, all proteins studied up-to-date, including ubiquitin, lysozyme, monellin and α-lactalbumin (see, for example, Alexandrescu et al. 1994, and references therein), as well as cytochrome c, carbonic anhydrase B and RBP (our unpublished

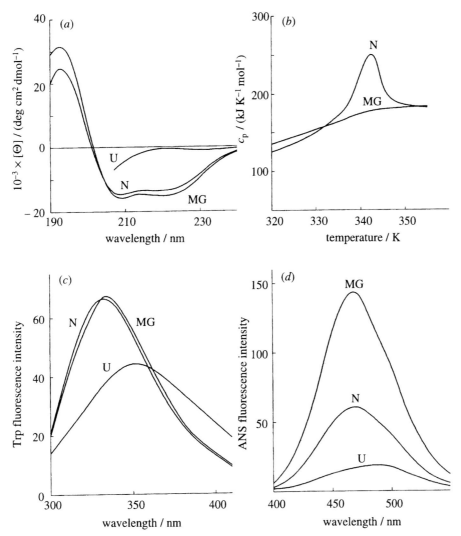

Figure 4. (*a*) Far UV CD spectra, (*b*) microcalorimetric recording, (*c*) Trp and (*d*) ANS fluorescence spectra for α-fetoprotein at pH 7.2 (N state), pH 3.1 (MG state) and in 9 M urea (U state). Adapted from Uversky *et al.* (1995).

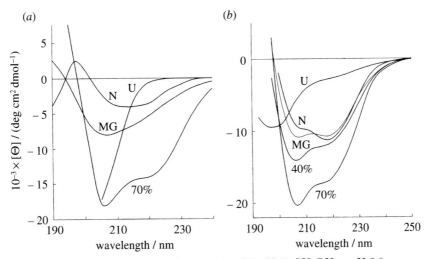

Figure 5. (*a*) Far UV CD spectra of bovine carbonic anhydrase B in 70% CH_3OH at pH 2.0 compared with those in N (pH 7.5, 0.5 M NaCL), MG (pH 3.6 in 10 mM phosphate buffer) and U (7 M GdmCl) states. (*b*) Far UV CD spectra of horse cytochrome *c* in 40 and 70% CH_3OH (pH 4.0, 0.5 M NaCl) compared with those in N (pH 4.0, 0.5 M NaCl), MG (pH 2.0, 0.5 M NaCl) and U (pH 2.0, water) states.

data), have much more pronounced far UV CD spectra in mixtures of water with large amounts of alcohols than in the native or in the molten globule states (see figure 5). In a similar way the addition of methanol

from 0 to 60% changes $[\Theta]_{220}$ of RBP from $+700$ at pH 7.5 (N state) and -6000 at pH 2.0 (MG state) to -9200 and -20000 deg cm^2 $dmol^{-1}$, correspondingly. This suggests that alcohol-denatured state of proteins may be substantially more helical than the N or the MG states. This conclusion is supported by NMR data suggesting alcohol-induced transitions of some β- or irregular chain regions of active proteins into α-helical state (see Alexandrescu *et al.* 1994, and references therein).

However, at smaller concentrations of methanol and moderately low pH, proteins have less pronounced far UV CD spectra. For example, for cytochrome c in 40% CH$_3$OH (figure 5b) or for RBP in 20% CH$_3$OH (data not shown), far UV CD spectra at pH 4 are close to those for the MG state of these proteins obtained without methanol at extreme pH (~ 2). The physical state of proteins at different alcohol concentrations requires more detailed investigation.

6. MOLTEN GLOBULES AND GENETIC DISEASES

As mentioned above, it is possible to obtain stable folding intermediates at physiological conditions by other means: some gene mutations can lead to a synthesis of mutant proteins which could not be completely folded under normal conditions. These mutations have been already reported (see, for example, Craig *et al.* 1985; Lim *et al.* 1992) and it was concluded that protein folding is stopped in these cases at the molten globule stage.

On the other hand, there are genetic diseases which can be provoked by point mutations in some proteins leading to their mislocation in a cell. This mechanism was well established for cystic fibroses (Yang *et al.* 1993) but is likely the case for some other diseases, for example for hypercholesterolemia and emphysema. This has led us (Bychkova & Ptitsyn 1994) to the assumption that mutations which cause genetic diseases by changing the intracellular pathway of proteins also inhibit the last stage of protein folding. These mutations may cause proteins to be trapped in the molten globule state which can either be associated with chaperones or aggregate. This prevents mutant proteins from normal trafficking in the cell and leads to mislocation and degradation of these proteins.

We thank A.E. Dujsekina and K.S. Vassilenko who took part in experiments on proteins in water–methanol mixtures. This research was supported in part by grants from the Human Frontier Scientific Foundation Program (Grant No. RG-331/93) and the Russian Foundation for Fundamental Investigations (Grant No. 93–04–6635).

REFERENCES

Abelev, G.I. 1993 Alpha-fetoprotein biology. *Sov. Sci. Rev. D. Physicochem. Biol.* **11**, 85–109.
Alexandrescu, A.T., Ng, Y.-L. & Dobson, C.M. 1994 Characterization of a trifluoroethanol-induced partially folded state of α-lactalbumin. *J. molec. Biol.* **235**, 587–599.
Baldwin, R.L. 1993 Pulsed H/D-exchange studies of folding intermediates. *Curr. Opin. Struct. Biol.* **3**, 84–91.
Bychkova, V.E. & Ptitsyn, O.B. 1993 The molten globule *in vitro* and *in vivo. Chemtracts Biochem. molec. Biol.* **4**, 133–163.
Bychkova, V.E. & Ptitsyn, O.B. 1995 Folding intermediates are involved in genetic disease? *FEBS Lett.* **359**, 6–8.

Bychkova, V.E. Pain, R.H. & Ptitsyn, O.B. 1988 The 'molten globule' state is involved in the translocation of proteins across membranes? *FEBS Lett.* **238**, 231–234.
Bychkova, V.E., Berni, R., Rossi, G.-L., Kutyshenko, V.P. & Ptitsyn, O.B. 1992 Retinol-binding protein is in the molten globule state at low pH. *Biochemistry* **31**, 7566–7571.
Craig, S., Hollecker, M., Creighton, T.E. & Pain, R.H. 1985 Single amino acid mutations block a late step in the folding of β-lactamase from *Staphylococcus aureus. J. molec. Biol.* **185**, 681–687.
Creighton, T.E. 1980 Kinetic study of protein unfolding and refolding using urea gradient electrophoresis. *J. molec. Biol.* **137**, 61–80.
Dolgikh, D.A., Gilmanshin, R.I., Brazhnikov, E.V. *et al.* 1981 α-Lactalbumin: compact state with fluctuating tertiary structure? *FEBS Lett.* **136**, 311–315.
Dolgikh, D.A., Kolomiets, A.P., Bolotina, I.A. & Ptitsyn, O.B. 1984 Molten globule state accumulates in carbonic anhydrase folding. *FEBS Lett.* **165**, 88–92.
Dolgikh, D.A., Abaturov, L.V., Bolotina, I.A. *et al.* 1985 Compact state of a protein molecule with pronounced small-scale mobility: bovine α-lactalbumin. *Eur. Biophys. J.* **13**, 109–121.
Elöve G.A., Chaffotte, A.F., Roder, H. & Goldberg, M.E. 1992 Early steps in cytochrome c folding probed by time-resolving circular dichroism and fluorescence spectroscopy. *Biochemistry* **31**, 6876–6883.
Fersht, A.R. 1993 Protein folding and stability: the pathway of folding of barnase. *FEBS Lett.* **325**, 5–16.
Gilmanshin, R.I. & Ptitsyn, O.B. 1987 An early intermediate of refolding α-lactalbumin forms within 20 ms. *FEBS Lett.* **223**, 327–329.
Hedrick, J.-P. & Hartl, F.U. 1993 Molecular chaperones functions of heat-shock proteins. *A. Rev. Biochem.* **62**, 349–384.
Hill, T.L. 1968 *Thermodynamics of small systems* (ed. W.A. Benjamin). New York: Wiley and Sons.
Jennings, P.A. & Wright, P.E. 1993 Formation of a molten globule intermediate early in the kinetic pathway of apomyoglobin. *Science, Wash.* **262**, 892–896.
Kawata, Y. & Hamaguchi, K. 1991 Use of fluorescence energy transfer to characterize the compactness of constant fragment of an immunoglobulin light chain in the early stage of folding. *Biochemistry* **30**, 4367–4373.
Kuwajima, K., Yamaya, H., Miwa, S., Sugai, S. & Nagamura, T.1987 Rapid formation of secondary structure framework in protein folding studied by stopped-flow circular dichroism. *FEBS Lett.* **221**, 115–118.
Kuwajima, K., Semisotnov, G.V. Finkelstein, A.V. Sugai, S. & Ptitsyn, O.B. 1993 Secondary structure of globular proteins at an early and the final stages in protein folding. *FEBS Lett.* **334**, 265–268.
Lim, W.A., Farrugio, D.C. & Sauer, R.T. 1992 Structural and energetic consequence of disruptive mutations in a protein core. *Biochemistry* **31**, 4324–4333.
Martin, J., Langer, T., Boteva, R., Schamel, A., Horwich, A.L. & Hartl, F.-U. 1991 Chaperonin-mediated protein folding at the surface of GroEL though a 'molten globule'-like intermediate. *Nature, Lond.* **352**, 36–42.
Palleros, D.R., Shi, L., Reid, K.L. & Fink, A.L. 1993 Three-state denaturation of DnaK induced by guanidine hydrochloride. Evidence for an expandable intermediate. *Biochemistry* **32**, 4314–4321.
Peng, Z. & Kim, P.S. 1994 A protein dissection study of a molten globule. *Biochemistry* **33**, 2136–2141.
Privalov, P.L. 1979 Stability of proteins. Small globular proteins. *Adv. Protein Chem.* **33**, 167–241.
Privalov, P.L. 1992 Physical basis of the stability of the folded conformations of proteins. In *Protein folding* (ed.

T.E. Creighton), pp. 83–126. New York: W.H. Freeman and Co.

Ptitsyn, O.B. 1973 Sequential mechanism of protein folding. *Dokl. Akad. Nauk SSSR* **210**, 1213–1215.

Ptitsyn, .O.B. 1987 Protein folding: hypothesis and experiments. *J. Prot. Chem.* **6**, 273–293.

Ptitsyn, O.B. 1992 The molten globule state. In *Protein folding* (ed. T.E. Creighton), pp. 243–300. New York: W.H. Freeman and Co.

Ptitsyn, O.B. 1994 Kinetic and equilibrium intermediates in protein folding. *Prot. Eng.* **7**, 593–596.

Ptitsyn, O.B. & Uversky, V.N. 1994 The molten globule is the third thermodynamical state of protein molecules. *FEBS Lett.* **341**, 15–18.

Radford, S.E., Dobson, C.M. & Evans, P.A. 1992 The folding of hen egg lysozyme involves partially structured intermediates and multiple pathways. *Nature, Lond.* **358**, 302–307.

Robson B. & Pain, R.H. 1976 The mechanism of folding of globular proteins: Equilibria and kinetics conformational transitions of penicillinase from *Staphylococcus aureus* involving a state of intermediate conformation. *Biochem. J.* **155**, 331–334.

Roder, H. & Elöve G.A. 1994 Early stages of protein folding. In *Mechanisms of protein folding* (ed. R.H. Pain), pp. 26–54. Oxford University Press.

Schatz, G. 1993 The protein import machinery of mitochondria. *Protein Sci.* **2**, 141–146.

Semisotnov, G.V. Rodionova, N.A., Kutyshenko, V.P., Ebert, B. Blank, J. & Ptitsyn, O.B. 1987 Sequential mechanism of refolding of carbonic anhydrase B. *FEBS Lett.* **224**, 9–13.

Semisotnov, G.V. Rodionova, N.A., Razgulyaev, O.I., Uversky, V.N., Gripas', A.F. & Gilmanshin, R.I. 1991 Study of the molten globule intermediate state by a hydrophobic fluorescent probe. *Biopolymers* **31**, 119–128.

Semisotnov, G.V., Kotova N.V., Kuwajima, K. *et al.* 1992 Monitoring protein folding pathway by stopped-flow X-ray scattering method. *Photon factory activity report* (National Laboratory for High Energy Physics, KEK, Japan) **10**, 366.

Shakhnovich, E.I. & Finkelstein, A.V. 1989 Theory of cooperative transitions in protein molecules. I. Why denaturation of globular protein is a first-order phase transition. *Biopolymers* **28**, 1667–1680.

Uversky, V.N. & Ptitsyn, O.B. 1994 'Partly-folded' state – a new equilibrium state of protein molecules: four-state guanidinium chloride-induced unfolding of β-lactamase at low temperature. *Biochemistry* **33**, 2782–2791.

Uversky, V.N. & Ptitsyn, O.B. 1995 Further evidence on the equilibrium 'pre-molten globule' state – a possible counterpart of first kinetic intermediate: four-state guanidinium chloride-induced unfolding of carbonic anhydrase B at low temperature. *J. molec. Biol.* (Submitted.)

Uversky, V.N., Semisotnov, G.V., Pain, R.H. & Ptitsyn, O.B. 1992 'All-or-none' mechanism of the molten globule unfolding. *FEBS Lett.* **314**, 89–92.

Uversky, V.N., Kirkitadze, M.D., Narizhneva, N.V., Potekhin, S.A. & Tomashevski, A.Yu. 1995 Structural properties of α-fetoprotein from human cord serum: protein molecule at low pH possesses all the properties of the molten globule. *FEBS Lett.* (Submitted.)

Van der Goot, F.G., Gonzales-Manes, J.M., Lakey, J.H. & Pattus, F. 1991 Molten globule membrane-insertion intermediate of the pore-forming domain of colicine-A. *Nature, Lond.* **354**, 408–410.

Van der Goot, F.G., Lakey, J.H. & Pattus, F. 1992 The molten globule intermediate for protein insertion or translocation through membranes. *Trends Cell Biol.* **2**, 343–348.

Yang, Y., Janich, S., Cohn, J.A. & Wilson, I.M. 1993 The common variant of cystic fibrosis transmembrane conductance regulator is recognized by hsp70 and degraded in a pre-Golgi nonlysosomal compartment. *Proc. natn. Acad. Sci. U.S.A.* **90**, 9480–9484.

Does the molten globule have a native-like tertiary fold?

ZHENG-YU PENG, LAWREN C. WU, BRENDA A. SCHULMAN
AND PETER S. KIM

Howard Hughes Medical Institute, Whitehead Institute for Biomedical Research, Department of Biology,
Massachusetts Institute of Technology, Nine Cambridge Center, Cambridge, Massachusetts 02142 U.S.A.

SUMMARY

One of the mysteries in protein folding is how folding intermediates direct a protein to its unique final structure. To address this question, we have studied the molten globule formed by the α-helical domain of α-lactalbumin (α-LA) and demonstrated that it has a native-like tertiary fold, even in the absence of rigid, extensive side chain packing. These studies suggest that the role of molten globule intermediates in protein folding is to maintain an approximate native backbone topology while still allowing minor structural rearrangements to occur.

A central issue in protein folding is to understand how a protein can fold quickly and efficiently to a unique native structure, despite the immense number of conformations accessible to the unfolded polypeptide (Levinthal 1968). This dilemma suggests that protein folding follows specific pathways, since an exhaustive search of all conformations is not possible on a physiological timescale. Thus, in order to understand protein folding, it is crucial to characterize the intermediates on these pathways.

Molten globules are partially folded forms of proteins proposed to be general intermediates in protein folding (Ptitsyn *et al.* 1990). Molten globules are characterized (table 1) by near-native levels of secondary structure but very little rigid, specific tertiary packing (for reviews, see Ptitsyn 1987, 1992; Kuwajima 1989; Christensen & Pain 1991). Equilibrium molten globules have spectroscopic properties and stabilities similar to that of early kinetic folding intermediates (Kuwajima *et al.* 1985; Ikeguchi *et al.* 1986; Jennings & Wright 1993). Yet, despite numerous studies, no high resolution structure of a molten globule is known, and the significance of molten globules to protein folding remains unclear.

The backbone topology of the molten globule (i.e., the relative orientations of secondary structure elements to one another) is a key unresolved issue. It has

been proposed that molten globules correspond to either (i) non-specific collapsed polypeptides or (ii) expanded native-like proteins (figure 1). Answering this question is of vital importance for understanding the role of molten globules in protein folding. If molten globules are non-specific collapsed polypeptides, then they would not contain much information for protein folding. On the other hand, if molten globules have a native-like topology, then they provide an approximate solution to the folding problem in which substantial information transfer has occurred.

α-Lactalbumin (α-LA) forms the best studied molten globule (Kuwajima *et al.* 1976, 1985; Nozaka *et al.* 1978; Dolgikh *et al.* 1981, 1985; Ikeguchi *et al.* 1986; Baum *et al.* 1989; Xie *et al.* 1991; Ewbank & Creighton 1991, 1993*a*, *b*; Alexandrescu *et al.* 1993; Creighton & Ewbank 1994; Peng & Kim 1994; Wu *et al.* 1995). The α-LA molten globule can be obtained under a variety of conditions, including low pH (the A-state) and in the presence of low concentrations of denaturants.

α-LA is a two-domain protein (figure 2). The helical domain consists of residues 1–37 and 85–123 and contains all four α-helices in α-LA. The β-sheet domain consists of residues 38–84 and contains a small antiparallel β-sheet and several irregular structures. α-LA has four disulphide bonds. Two of the disulphide bonds (6–120 and 28–111) are in the α-helical domain, one (61–77) is in the β-sheet domain, and one (73–91) connects the two domains.

We decided to study the molten globule of α-LA by 'protein dissection,' removing parts of the protein molecule deemed extraneous (Peng & Kim 1994). The resulting molecule is called α-Domain and consists exclusively of residues from the helical domain of α-LA (figure 2; for simplicity, we use the same numbering system as used for α-LA). We focused on the α-helical domain for several reasons. First, the helical content of the α-LA molten globule is similar to that of native α-

Table 1. *Common characteristics of molten globules*

substantial level of secondary structure, often comparable with that of the native protein

absence of well-defined side-chain packing

lack of a cooperative thermal unfolding transition

compactness

Phil. Trans. R. Soc. Lond. B (1995) **348**, 43–47
Printed in Great Britain

43

LA, as judged by circular dichroism (CD) studies (Kuwajima *et al.* 1976; Nozaka *et al.* 1978; Dolgikh *et al.* 1981, 1985). Second, nuclear magnetic resonance (NMR) studies show that some of the native helices, but little of the β-sheet domain, are structured in the molten globule of α-LA (Baum *et al.* 1989; Alexandrescu *et al.* 1993; Chyan *et al.* 1993). Finally, the helical domain of hen egg-white lysozyme, a protein that is structurally homologous to α-LA, has been shown to fold prior to the β-sheet domain (Miranker *et al.* 1991, 1993; Radford *et al.* 1992).

α-Domain (figure 2) with the native disulphide pairings between residues 28–111 and 6–120 (called α-Domain^{ox}) exhibits the characteristics of a molten globule and is a good model system for molten globule studies (Peng & Kim 1994). α-Domain^{ox}is a compact monomer at low concentrations (pH 8.5, no additional salt), contains substantial helical secondary structure, and lacks rigid side-chain packing. α-Domain^{ox} and native α-LA have approximately the same number of residues in an α-helical conformation, as measured by far-UV CD spectroscopy. The near-UV CD and one-dimensional proton NMR spectra of α-Domain^{ox} are very similar to that of the A-state molten globule of α-LA. α-Domain^{ox} also exhibits a non-cooperative thermal transition similar to that of the molten globule of intact α-LA.

To investigate the backbone topology of α-Domain, we performed equilibrium disulphide exchange experiments (Peng & Kim 1994). The cysteines in α-Domain are located in disparate parts of the molecule. Thus, the relative populations of disulphide-bonded species reflect the probability of forming different backbone topologies. If disulphide bond formation is random, then the relative populations can be calculated based on a random-walk model (Kauzmann 1959), which predicts that only 7% of the fully oxidized molecules should have the native disulphide bonds. Strikingly, we observe that ∼ 90% of the fully oxidized molecules have native disulphide bonds (i.e. α-Domain^{ox}) under native conditions (figure 3). In contrast, under denaturing conditions, ∼ 8% of the molecules were found to have the native disulphide bonds, in agreement with the value predicted by the random walk

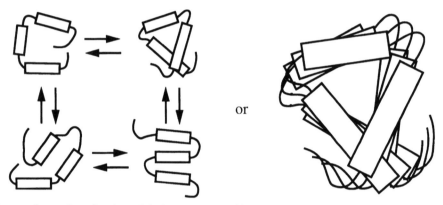

Figure 1. A central question of molten globule structure and its role in protein folding is whether the molten globule has a native-like tertiary fold.

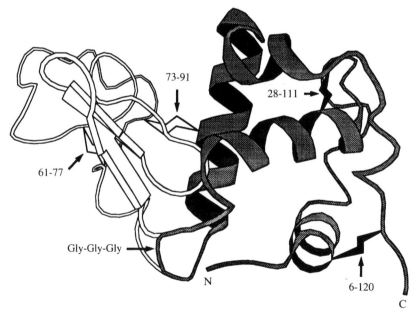

Figure 2. Schematic representation of α-LA. The recombinant α-Domain (shaded) consists of residues 1–39 and 81–123 of human α-LA connected by a short linker of three glycines and preceded by an N-terminal methionine. The two disulphide bonds in α-Domain (6–120 and 28–111) are shown in black. Cys 91, which forms an inter-domain disulphide bond in α-LA, has been changed to alanine to avoid unwanted thiol-disulphide reactions.

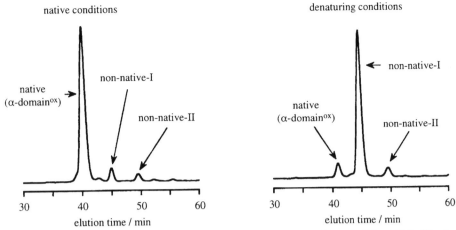

Figure 3. HPLC analysis of the disulphide exchange in α-Domain at pH 8.5 (Peng & Kim 1994). The 'native' isomer (also called α-Domain^ox) contains disulphide bonds 28–111 and 6–120. The 'non-native-I' isomer contains disulphide bonds 6–28 and 111–120, and the 'non-native-II' isomer contains disulphide bonds 6–111 and 28–120. Under native conditions, the ratio of native : non-native-I : non-native-II is 90:6:4. Under denaturing conditions (6 M GuHCl), the ratio of native : non-native-I : non-native-II is 8:85:7. For comparison, the ratio predicted for a random walk model is 7:88:5, where the probability of forming an intramolecular disulphide bond is proportional to $n^{-3/2}$ and $n-1$ is the number of intervening non-cysteine residues in the loop.

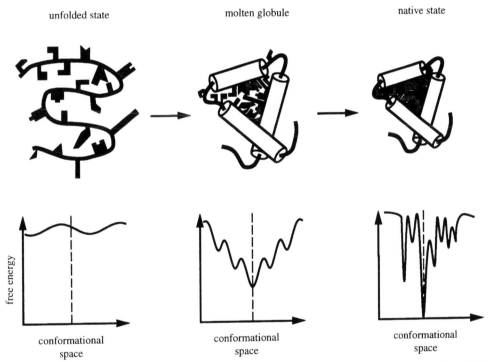

Figure 4. A two-step model of protein folding. The hypothetical structures of a protein in the unfolded, molten globule, and native states are shown along with the free energy landscapes at different stages of the folding reaction.

model. Circular dichroism studies indicate that α-Domain^ox has substantially more helical secondary structure than reduced α-Domain (Peng & Kim 1994). In contrast, both non-native disulphide bond isomers ((6–28; 111–120) and (6–111; 28–120)) contain significantly less helical secondary structure than reduced α-Domain. Taken together, these data indicate that the polypeptide backbone of α-Domain prefers a native-like topology and that non-native topologies imposed by non-native disulphide bonds are inconsistent with the high level of secondary structure found in the molten globule of intact α-LA.

Disulphide exchange studies of the molten globule of the entire α-LA molecule with all eight cysteines intact

failed to show a strong preference for the species with native disulphide pairings (Ewbank & Creighton 1991, 1993a). Instead, many disulphide bond isomers were significantly populated, and only average properties could be examined, since individual disulphide species could not be studied separately. These results were interpreted to indicate that molten globules are actually non-specific collapsed polypeptides, in apparent contradiction to the studies of α-Domain. Two explanations of this discrepancy are apparent.

First, it is possible that the three glycine residues in α-Domain, substituting for 41 residues of the β-sheet domain, constrain the flexibility of the α-Domain backbone, thereby providing a bias toward a topology with native disulphide pairings. Alternatively, the β-sheet domain may be unstructured in the molten globule of intact α-LA. If this were the case, then those α-LA molecules with native disulphide pairings in the helical domain would be spread among three equally probable β-sheet domain and interdomain disulphide pairings. In addition, interdomain disulphide exchange would further obscure the preference of the helical domain for native disulphide pairings. A rough calculation of the expected disulphide populations can be made, assuming, as a crude approximation, that those species of α-LA with native disulphide pairings in the α-helical domain are favoured twenty-fold over all other species. Then, the populations of α-LA molecules with native disulphide bonds in the α-helical domain would be spread over three peaks, each containing ∼ 12 % of the population, with the rest of the α-LA population contained in the other 102 disulphide isomers.

Recent experimental results indicate that the molten globule form of intact α-LA has a bipartite structure (Wu *et al.* 1995). The α-helical domain strongly favours the native backbone topology, while the β-sheet domain is largely unstructured. This finding provides a likely resolution to the apparent discrepancy between the studies of α-Domain (Peng & Kim 1994) and intact α-LA (Ewbank & Creighton 1991, 1993*a*) molten globules, as outlined above. In addition, these results demonstrate that molten globule properties need not encompass the entire polypeptide chain and can be achieved independently by individual domains.

Molten globules can form very quickly (typically ≤ 20 ms; see, for example, Kuwajima *et al.* 1987; Gilmanshin & Ptitsyn 1987). Late folding intermediates containing extensive tertiary interactions, including the so-called 'highly ordered molten globule' (Redfield *et al.* 1994; Feng *et al.* 1994), are known to be native-like. Our results suggest that even early folding intermediates, such as molten globules (in the traditional sense), have a native-like tertiary fold, providing a quick and approximate solution to the Levinthal paradox.

Our results suggest a two-step model for protein folding in which early formation of the molten globule achieves much of the information transfer from one- to three-dimensions (figure 4). Thus, the role of molten globules in protein folding is to maintain an approximate native-like structure, thereby greatly decreasing the conformational space to be searched by the polypeptide chain and preventing global misfoldings. The subsequent search for a unique folded conformation is facilitated by the flexibility of molten globules, which reduces the energy barriers for side chain rearrangements.

How do molten globules achieve the native tertiary fold? One extreme is that molten globules contain specific tertiary interactions that are not detectable by near-UV CD and NMR studies. The other extreme is that more global features, such as the pattern of hydrophilic and hydrophobic residues, side-chain volumes, and local secondary structure propensities largely determine the tertiary fold of a protein. Further experiments should resolve this important question.

This work was supported by the NIH (GM-41307).

REFERENCES

Alexandrescu, A.T., Evans, P.A., Pitkeathly, M., Baum, J. & Dobson, C.M. 1993 Structure and dynamics of the acid-denatured molten globule state of α-lactalbumin: a two-dimensional NMR study. *Biochemistry* **32**, 1707–1718.

Baum, J., Dobson, C.M., Evans, P.A. & Hanley, C. 1989 Characterization of a partly folded protein by NMR methods: Studies on the molten globule state of guinea pig α-lactalbumin. *Biochemistry* **28**, 7–13.

Christensen, H. & Pain, R.H. 1991 Molten globule intermediates and protein folding. *Eur. Biophys. J.* **19**, 221–229.

Chyan, C.L., Wormald, C., Dobson, C.M., Evans, P.A. & Baum, J. 1993 Structure and stability of the molten globule state of guinea-pig alpha-lactalbumin: a hydrogen exchange study. *Biochemistry* **32**, 5681–5691.

Creighton, T.E. & Ewbank, J.J. 1994 Disulfide rearranged molten globule state of α-lactalbumin. *Biochemistry* **33**, 1534–1538.

Dolgikh, D.A., Abaturov, L.V., Bolotina, I.A. *et al.* 1985 Compact state of a protein molecule with pronounced small-scale mobility. *Eur. Biophys. J.* **13**, 109–121.

Dolgikh, D.A., Gilmanshin, R.I., Brazhnikov, E.V. *et al.* 1981 α-Lactalbumin: compact state with fluctuating tertiary structure. *FEBS Lett.* **136**, 311–315.

Ewbank, J.J. & Creighton, T.E. 1991 The molten globule protein conformation probed by disulphide bonds. *Nature, Lond.* **350**, 518–520.

Ewbank, J.J. & Creighton, T.E. 1993*a* Pathway of disulfide-coupled unfolding and folding of bovine alpha-lactalbumin. *Biochemistry* **32**, 3677–3693.

Ewbank, J.J. & Creighton, T.E. 1993*b* Structural characterization of the disulfide folding intermediates of bovine alpha-lactalbumin. *Biochemistry* **32**, 3694–3707.

Feng, Y.-q., Sligar, S.G. & Wand, A.J. 1994 Solution Structure of Apo-cytochrome b562. *Nature Struct. Biol.* **1**, 30–35.

Gilmanshin, R.I. & Ptitsyn, O.B. 1987 An early intermediate of refolding α-lactalbumin forms within 20 ms. *FEBS Lett.* **223**, 327–329.

Ikeguchi, M., Kuwajima, K., Mitani, M. & Sugai, S. 1986 Evidence for identity between the equilibrium unfolding intermediate and a transient folding intermediate: A comparitive study of the folding reactions of α-lactalbumin and lysozyme. *Biochemistry* **25**, 6965–6972.

Jennings, P.A. & Wright, P.E. 1993 Formation of a molten globule intermediate early in the kinetic folding pathway of apo-myoglobin. *Science, Wash.* **262**, 892–896.

Kauzmann, W. 1959 Relative probabilities of isomers in cystine-containing randomly coiled polypeptides. In *Sulfur in proteins* (ed. R. Benesch, R.E. Benesch, P.D. Boyer *et al.*), pp. 93–108. New York: Academic Press.

Kuwajima, K. 1989 The molten globule state as a clue for understanding the folding and cooperativity of globular-protein structure. *Proteins Struct. Funct. Genet.* **6**, 87–103.

Kuwajima, K., Hiraoka, Y., Ikeguchi, M. & Sugai, S. 1985 Comparison of the transient folding intermediates in lysozyme and α-lactalbumin. *Biochemistry* **24**, 874–881.

Kuwajima, K., Nitta, K., Yoneyama, M. & Sugai, S. 1976 Three-state denaturation of α-lactalbumin by guanidine hydrochloride. *J. molec. Biol.* **106**, 359–373.

Kuwajima, K., Yamaya, H., Miwa, S., Sugai, S. & Nagamura, T. 1987 Rapid formation of secondary structure framework in protein folding studied by stopped-flow circular dichroism. *FEBS Lett.* **221**, 115–118.

Levinthal, C. 1968 Are there pathways for protein folding? *J. Chim. Phys.* **85**, 44–45.

Miranker, A., Radford, S.E., Karplus, M. & Dobson, C.M. 1991 Demonstration by NMR of folding domains in lysozyme. *Nature, Lond.* **349**, 633–636.

Miranker, A., Robinson, C.V., Radford, S.E., Alpin, R.T. & Dobson, C.M. 1993 Detection of transient protein folding populations by mass spectrometry. *Science, Wash.* **262**, 896–900.

Nozaka, M., Kuwajima, K., Nitta, K. & Sugai, S. 1978 Detection and characterization of the intermediate on the folding pathway of human α-lactalbumin. *Biochemistry* **17**, 3753–3758.

Peng, Z.-y. & Kim, P.S. 1994 A protein dissection study of a molten globule. *Biochemistry* **33**, 2136–2141.

Ptitsyn, O.B. 1987 Protein folding: hypotheses and experiments. *J. Protein Chem.* **6**, 273–293.

Ptitsyn, O.B. 1992 The molten globule state. In *Protein folding* (ed. T.E. Creighton), pp. 243–300. New York: W.H. Freeman and Co.

Ptitsyn, O.B., Pain, R.H., Semisotnov, G.V., Zerovnik, E. & Razgulyaev, O.I. 1990 Evidence for a molten globule state as a general intermediate in protein folding. *FEBS Lett.* **262**, 20–24.

Radford, S.E., Dobson, C.M. & Evans, P.A. 1992 The folding of hen lysozyme involves partially structured intermediates and multiple pathways. *Nature, Lond.* **358**, 302–307.

Redfield, C., Smith, R.A.G. & Dobson, C.M. 1994 Structural characterization of a highly-ordered 'molten globule' at low pH. *Nature Struct. Biol.* **1**, 23–29.

Wu, L.C., Peng, Z.-y. & Kim, P.S. 1995 Bipartite structure of the α-lactalbumin molten globule. *Nature Struct. Biol.* **2**, 281–286.

Xie, D., Bhakuni, V. & Freire, E. 1991 Calorimetric determination of the energetics of the molten globule intermediate in protein folding: Apo-α-lactalbumin. *Biochemistry* **30**, 10673–10678.

Investigation of protein unfolding and stability by computer simulation

W. F. VAN GUNSTEREN, P. H. HÜNENBERGER, H. KOVACS, A. E. MARK
AND C. A. SCHIFFER

Laboratory of Physical Chemistry, ETH Zürich, 8092 Zürich, Switzerland

SUMMARY

Structural, dynamic and energetic properties of proteins in solution can be studied in atomic detail by molecular dynamics computer simulation. Protein unfolding can be caused by a variety of driving forces induced in different ways: increased temperature or pressure, change of solvent composition, or protein amino acid mutation. The stability and unfolding of four different proteins (bovine pancreatic trypsin inhibitor, hen egg white lysozyme, the surfactant protein C and the DNA-binding domain of the 434 repressor) have been studied by applying the afore-mentioned driving forces and also to some artificial forces. The results give a picture of protein (in)stability and possible unfolding pathways, and are compared to experimental data where possible.

1. INTRODUCTION

In the biologically active form, proteins generally assume a specific, folded conformation. Knowledge of native protein conformations has been derived from a variety of experiments. Three-dimensional structural models have been derived from X-ray diffraction of crystals and from multi-dimensional nuclear magnetic resonance (NMR) experiments. By comparison, very little is known about the denatured state of a protein or about the processes of denaturation and folding. Experimentally, it is very difficult to obtain structural information about these processes because of the atomic length scale and the short timescale of these processes. Partial structural information on intermediate states of the folding or unfolding pathways of proteins may be obtained by monitoring different spectroscopic properties of amino acid side chains or particular protein atoms or labelling groups as a function of time after induction of folding or denaturation (Dobson *et al.* 1994). Partly denatured states may be trapped and studied in atomic detail (Neri *et al.* 1992 *b*).

In view of the difficulty of the study of protein folding and denaturation by experimental means, one may turn to the method of computer simulation of these processes as a means of elucidating their nature in atomic detail. Theoretical methods to predict protein properties can be classified as follows.

1. Methods to assess the correctness of a given protein structure (see, for examples, Novotny *et al.* 1984; Jones *et al.* 1992; Lüthy *et al.* 1992; Maiorov & Crippen 1992; Ouzounis *et al.* 1993; Kocher *et al.* 1994), and for the prediction of protein structure (see, for examples, Dill 1985, 1990; Covell & Jernigan 1990; Unger & Moult 1993; Kolinski & Skolnick 1994).

2. Methods to determine the relative stability of protein mutants or different protein conformations in terms of relative free energies. It has been shown (Shi Yun-yu *et al.* 1993) that free-energy simulation techniques cannot currently be used to reliably predict protein stability.

3. Methods to simulate the process of protein folding or denaturation. Here, one thinks primarily of molecular dynamics (MD) simulation or of Monte Carlo (MC) simulation in cases where the MC step and acceptance criterion are of a physical nature (see, for examples, Levitt & Warshel 1975; Sali *et al.* 1994).

The simulation of protein folding is severely limited by two factors.

1. The size of the conformational space to be searched for the native state of low free energy.

2. The lack of structural information with respect to the denatured (possibly unfolded) state of a protein, which may be characterized by manifold conformations.

To reduce the size of the conformational space, very simple protein models (e.g. models using one interaction site per amino acid residue) are used (Seetharamulu & Crippen 1991), or the accessible protein conformations can be restricted to those fitting on a regular spatial lattice (Covell & Jernigan 1990). The lack of knowledge of the structures that constitute the denatured state can be compensated by repeating the folding process starting from different random conformations. However, owing to the limited accuracy of very simple protein models and the heuristic nature of most search methods used, reliable simulation of protein folding is still out of reach.

The process of protein destabilization or denaturation can be simulated much more easily on a computer, because atomic detail of the initial conformation is known for many proteins; but even these simulations are limited.

1. Because the process of protein unfolding (and folding) is driven by the interplay between protein–

Phil. Trans. R. Soc. Lond. B (1995) **348**, 49–59
Printed in Great Britain

49

© 1995 The Royal Society and the authors

protein, protein–solvent and solvent–solvent interactions at a given temperature, pressure, pH, etc., the inclusion of explicit solvent molecules in an unfolding simulation is mandatory to obtain useful results.

2. Simulation of a protein in explicit solvent requires substantial computing power, which currently limits the timescale of such simulations to nanoseconds.

3. The interatomic interaction function used should give the correct relative strength of protein–protein, protein–solvent and solvent–solvent interactions, a feature which is not necessarily present when, for example, protein and solvent force fields of different type or parameterization are combined.

In this paper we study protein stability and the onset of denaturation by MD simulation under denaturing conditions in the presence of explicit solvent molecules.

The unfolding of a protein can be induced by different driving forces. These include: amino acid mutation (e.g. the change of cysteine to alanine residues); increase in temperature; increase in pressure; change in solvent type (e.g. from a polar one, such as water, to a less polar one, such as chloroform); and change in the pH or ionic strength of the solution containing the protein.

Artificial driving forces, such as the application of a constant radial (from the protein centre of mass) force or a constant radial temperature gradient, have also been used (Hao *et al.* 1993; Hünenberger *et al.* 1995), but have been shown to introduce considerable bias to the unfolding process (Hünenberger *et al.* 1995).

Technically, the unfolding pathway of a protein can be studied by two different types of simulation.

1. By performing equilibrium simulations of the protein in solution under different conditions of temperature, pressure, pH, etc., or for different mutants and solvent composition. In this case the equilibrium properties averaged over the different equilibrium simulations are compared, and it is assumed that the state points at which the protein is simulated are representative for points along its denaturation pathway.

2. By performing, in parallel, equilibrium and non-equilibrium simulations starting from a single (equilibrated) structure. Equilibrium of the system is perturbed by switching on one of the mentioned driving forces and its relaxation towards the (new) equilibrium, which may be a denatured state, is monitored. In this case the change of properties over time, and not the averaged properties, is analysed.

Both types of simulation give information about protein stability. A protein would be considered more stable if it displays structural integrity alongside a small degree of fluctuation at equilibrium under various (extreme) external conditions (temperature, pressure, etc.). When performing non-equilibrium simulations the protein would be considered more stable the slower it reacts or gives in to the perturbing forces.

A variety of structural properties and some energetic properties can be monitored (out of equilibrium) or averaged (in equilibrium) and subsequently compared between simulations under different conditions or with experimental data. The radius of gyration of the protein can be compared to light scattering data and changes in the polar and non-polar solvent accessible area of the protein can be used to estimate the change in heat capacity upon unfolding. A changing amount of helicity may be compared with circular dichroism data. Atom–atom distances may be compared to nuclear overhauser enhancement (NOE) intensities obtained by NMR spectroscopy. The deviation of the protein structure from its crystalline X-ray structure may be analysed. Presence or absence of hydrogen bonds may be correlated with data from hydrogen–deuterium exchange experiments. Protein energies may be compared with experimental melting temperatures. Yet, because the experimental data on protein properties at different points along the pathways of denaturation are scarce, the reliability of the denaturation simulations cannot be established by a comparison of simulated properties with these experimental data. The confidence in the denaturation simulations will largely stem from the correct reproduction of native protein properties under native conditions using the same interaction function and similar simulation protocols.

Simulation studies of protein stability and denaturation in explicitly simulated solvents have only recently become possible with the increasing power of computers. Mark & van Gunsteren (1992) studied thermal unfolding of hen egg white (HEW) lysozyme using non-equilibrium MD at high temperature (500 K). Daggett & Levitt (1992, 1993) studied thermal unfolding of bovine pancreatic trypsin inhibitor (BPTI) and its reduced form in equilibrium at different temperatures (423 K, 498 K). A similar type of study was performed by Daggett (1993) for the C-terminal fragment (CTF) of the L7/L12 ribosomal protein, and by Caflisch & Karplus (1994) for the protein barnase. Tirado-Rives & Jorgensen (1993) studied the effect of a change in pH by comparing simulations of myoglobin in which histidine residues were differently protonated. Simulations at high pressure have been reported by Kitchen *et al.* (1992) and Brunne & van Gunsteren (1993) for BPTI, and by Hünenberger *et al.* (1995) for HEW lysozyme. The effect of different cysteine to alanine mutations on the stability of BPTI at room temperature has been studied by Schiffer & van Gunsteren (1995). The effect of a change of solvent has been studied for the surfactant lipoprotein C by Kovacs *et al.* (1995) and for the DNA binding domain of the 434 repressor by Schiffer *et al.* (1995).

Here we briefly review the four last-mentioned studies to obtain a general picture of protein stability and the onset of denaturation under the influence of different driving forces. Because the computational set-up and details of analysis have been reported elsewhere, we only describe the simulations and results in a global manner.

2. STRUCTURAL STABILITY OF DISULPHIDE MUTANTS OF BPTI

Experimentally, the folding pathway of BPTI has been studied extensively (Weissman & Kim 1991; Darby *et al.* 1992; Goldenberg 1992). In its native form (see figure 1) BPTI has three disulphide bonds (5–55, 14–38, 30–51). The three forms of BPTI having two native disulphide bridges maintain the structure and function of the native three disulphide species. However, the three forms of BPTI having only one native disulphide bridge show different structural properties. Using NMR it has been found that the (5–55) species is folded in the native conformation. It can inhibit trypsin. The (14–38) species is not likely to adopt a well defined structure because this disulphide bond lies at the surface of the protein (see figure 1), and is the first to break and the last to form. The (30–51) species maintains native-like structure for residues 19–36 (β-sheet) and 42–56 (α-helix), the N-terminal part and the loop between residues 37 and 41 being disordered (van Mierlo *et al.* 1992). The fully reduced BPTI shows no stable secondary or tertiary structure.

In the light of this experimental information we decided to study the relative structural stability of the native form, the (5–55) species, the (30–51) species and the no-disulphide (all Cys to Ala) mutant of BPTI using non-equilibrium MD simulations. The four species were each simulated in aqueous solution at room temperature (at 300 K for 150 ps, then at 320 K for 100 ps), starting from the same equilibrated solution structure of native BPTI. Each system consisted of one BPTI molecule in a periodic truncated octahedron containing 2371 water molecules. The Gronigen molecular simulation (GROMOS) protein force field (van Gunsteren & Berendsen 1987) and the SPC/E water model (Berendsen *et al.* 1987) were used. Computational details are given in Schiffer & van Gunsteren (1995).

Figure 2 shows a comparison of the common starting structure with the structure after 250 ps of MD simulation of the four different species. The native BPTI (figure 2*a*) maintains its native structure, apart from a few N- and C-terminal residues which are also found to be disordered experimentally. The (5–55) species (figure 2*b*) essentially maintains the native structure, although slight deformations in the loop regions are observed. The (30–51) species (figure 2*c*) shows structural changes for residues 1–17 and residues 38–43 with respect to the rest of the molecule, which corresponds directly to the part of this mutant observed to be disordered experimentally. The non-disulphide species (figure 2*d*) shows much more deformation over the 250 ps simulation period.

These results show that parallel MD simulations of different protein mutants starting from the same initial structure may be used to obtain indications of the relative structural stability of the protein mutants.

3. STRUCTURAL STABILITY OF HEW LYSOZYME UNDER CONDITIONS OF HIGH TEMPERATURE OR PRESSURE

The folding and unfolding of HEW lysozyme has been studied extensively by a great variety of experimental methods (Dobson *et al.* 1994). Most studies concern temperature or denaturant-induced unfolding, but pressure-induced unfolding data have also been reported (Samarasinghe *et al.* 1992). In its native state (see figure 3*a*) HEW lysozyme is characterized by two structural domains: one consisting primarily of α-helix (α-domain) and the other dominated by a section of triple stranded β-sheet (β-domain). It has been shown that the two domains behave differently under different denaturing conditions, and are largely independent of each other during folding and unfolding processes. These factors make lysozyme a particularly interesting system in which to study such processes. At pressures up to 5 kbar, a low pH of 3.9 and a temperature of 342 K (which is close to the thermal denaturation temperature of about 350 K at 1 bar and pH of 7.0) the maximum degree of denaturation observed by NMR was about 53%. In X-ray diffraction studies of lysozyme crystals at high pressure (Kundrot & Richards 1987) deformation was observed primarily in the α-domain. In contrast, hydrogen exchange studies during refolding from a denaturant-induced unfolded state indicate

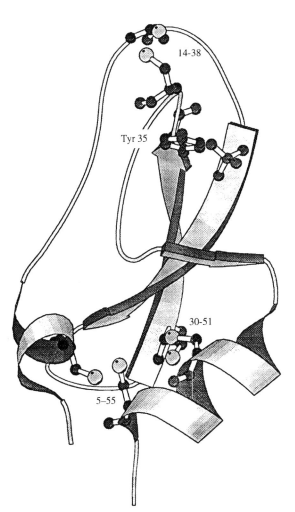

Figure 1. Schematic drawing of the crystal structure of BPTI showing the location of the three native disulphide bridges, the helices and the β-sheet.

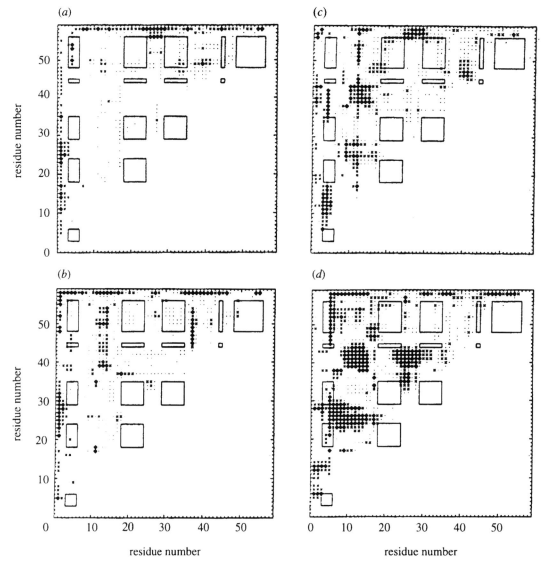

Figure 2. Residue–residue matrices showing the change in C_α–C_α atom distances for four BPTI species occurring in four MD simulations of 250 ps length at room temperature starting from an equilibrated structure of native BPTI in aqueous solution. If a C_α–C_α distance has changed less than 0.1 nm over the 250 ps, the residue–residue matrix displays a blank; if the distance has changed by 0.1–0.2 nm, it shows a dot; a change of 0.2–0.3 nm is indicated by a cross; changes of more than 0.3 nm are indicated by a rhombus. (*a*) native BPTI; (*b*) (5–55) species; (*c*) (30–51) species; (*d*) no disulphide species. Secondary structure elements and their relative position in the amino acid sequence are indicated by boxes.

preferential stabilization of the α-domain. The helices in the α-domain (see figure 3*a*) are protected against hydrogen exchange from the onset of folding, whereas the β-sheet amide hydrogens are only protected in the final stages of denaturant-induced refolding (Radford & Dobson, this volume).

In view of this wealth of experimental information, we decided to study the denaturation of HEW lysozyme in aqueous solution by non-equilibrium MD simulation using different driving forces. The system consisted of one HEW lysozyme molecule in a rectangular periodic box of $4.9 \times 5.3 \times 6.8$ nm³ containing 5345 water molecules. The GROMOS protein force field (van Gunsteren & Berendsen 1987) and the SPC water model (Berendsen *et al.* 1981) were used: computational details are given in Mark & van Gunsteren (1992) and Hünenberger *et al.* (1995).

Figure 3*a* shows the solution structure of HEW lysozyme equilibrated for 50 ps at 300 K and 1 bar

pressure; this should be compared with the structure obtained after an additional 210 ps of MD simulation at 342 K and a pressure of 10 kbar (see figure 3*c*). Within 210 ps, only slight denaturation is observed: the D helix unfolds. A comparison of the compressibility of the two domains of HEW lysozyme shows that the β-domain is essentially incompressible whereas the α-domain and the interdomain hinge region contract under pressure, a result which agrees with an X-ray diffraction study of HEW lysozyme crystals under 1 kbar pressure (Kundrot & Richards 1987).

Increasing the temperature to 500 K after 50 ps MD simulation at 300 K and 1 bar pressure, results in rapid unfolding of lysozyme. Figure 3*b* shows the structure after 120 ps of high temperature motion and figure 3*d* that which is obtained after 180 ps. Figure 4 shows the presence of different types of (secondary) structure elements as a function of time. With an increase in temperature from 300 to 500 K at 50 ps, a differential

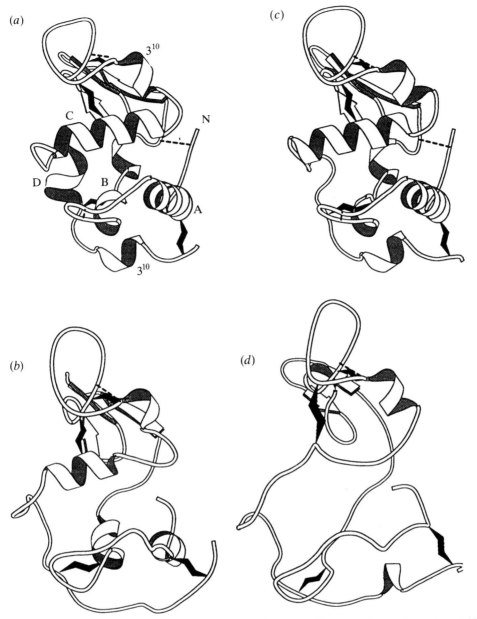

Figure 3. Ribbon models for the different structures of HEW lysozyme after: (*a*) 50 ps equilibration at 300 K and 1 bar (*b*) 120 ps at 500 K (*c*) 210 ps at 342 K and 10 kbar (*d*) 180 ps at 500 K.

denaturation of the different parts of the molecule is observed. The B helix breaks down, but refolds shortly after 110 ps. The D helix and terminal 3^{10} helix do not seem to be very stable. The C helix and nearby 3^{10} helix resist denaturation for the longest time, followed by a part of the B helix and then the A helix. The β-sheet is very resistant to denaturation. The fluctuating nature of the intra-protein hydrogen bonds displayed in figure 4 shows that different secondary structure elements may be formed along the (un-)folding pathway. The occurrence of stable hydrogen bonds outside secondary structure elements shown in figure 4 suggests that protection of a backbone amide hydrogen may result from a transient (un-)folding conformation and need not necessarily result from a direct involvement in secondary structure.

These results of pressure- and temperature-induced denaturation of HEW lysozyme show that computer simulation of these processes can be used to obtain insight, at the atomic level, into the differential structural stability of parts of a protein and with respect to the initial stages of the denaturation process.

4. STRUCTURAL STABILITY OF THE SURFACTANT PROTEIN C IN CHLOROFORM, METHANOL AND WATER

The stability of a folded protein will depend not only on its amino acid sequence, the temperature and pressure, but also on the nature of its solvent or membrane environment. The stability of an α-helix is especially sensitive to its environment. For example, trifluoroethanol is known to induce helicity of peptides in solution, and peptides with stretches of hydrophobic residues are assumed to adopt a helical conformation in a lipid environment. In these cases the stability of a given helix will depend on an energetic balance between interactions between spatially neighbouring amino acids in the polypeptide chain and interactions

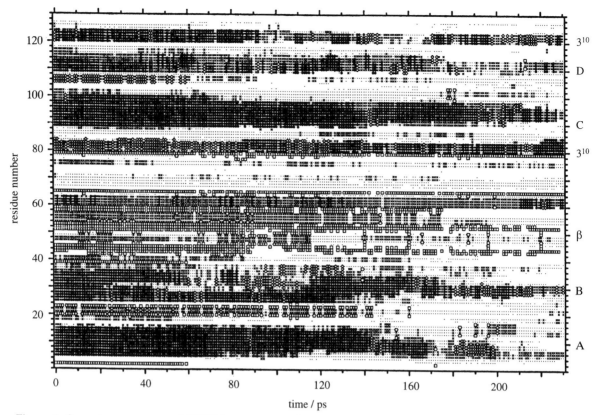

time / ps

Figure 4. Secondary structure of HEW lysozyme as a function of time. Until 50 ps the temperature is 300 K, afterwards it is 500 K. α-helix: $\bullet$; β-bridge or sheet: $\square$; 3^{10}-helix: $\diamond$; π-helix: $\blacklozenge$; hydrogen bonded turn: $\times$; bend: $\cdot$; all according to the DSSP program (Kabsch & Sander 1983).

Table 1. *Physical properties of the solvents at 298 K (Lide 1993)*

	dipole moment	density	molar volume	dielectric constant	viscosity	polarizability $(\alpha_0/4\pi\epsilon_0)$
	Debye	(g cm^{-3})	(cm^3 mol^{-1})		cP	10^{-24} cm^3
chloroform (Dietz *et al.* 1984, 1985)	1.04 (gas) 1.10 (MD)	1.480 (exp) 1.484 (MD)	80.7	4.7	0.54	9.5
methanol (Stouten 1989)	1.70 (gas) 2.32 (MD)	0.787 (exp) 0.791 (MD)	40.7	32.7	0.54	3.3
water (Berendsen *et al.* 1981)	1.85 (gas) 2.27 (MD)	0.997 (exp) 1.000 (MD)	18.1	78.4	0.89	1.5

to the solvent environment. Stability also depends on the rearrangement of the solvent medium that is induced by the presence of the protein. An interesting case is presented by the lung surfactant protein C (SP-C), which is an essential component of pulmonary surfactant, a mixture composed mainly of phospholipids and a few specific proteins which reduces the surface tension at the alveolar air–liquid interface in the lung thereby preventing alveolar collapse at end expiration (Johansson *et al.* 1994a). The 35 residue SP-C contains stretches of seven (15–21) and four (23–26) consecutive valine residues. It maintains an α-helical form between residues 9 and 34 in a 1:2 mixture chloroform:methanol containing 5% 0.1 M HCl. The eight N-terminal residues lack a stable regular secondary structure (Johansson *et al.* 1994b).

To investigate the effect of environment suggested by this experimental information we studied the relative stability of SP-C in three different pure solvents – chloroform, methanol and water – by equilibrium MD simulations. The use of pure solvents avoids the difficulty of obtaining a proper equilibration of a mixed solvent within the simulation timespan and enables a more straightforward comparison of protein–solvent interactions. The physical properties of the three solvents are quite different (see table 1), and it was ensured that the solvent models used in the simulations correctly reproduced the pure liquid properties of the solvents. The GROMOS protein force field (van Gunsteren & Berendsen 1987) with the correction indicated by Mark *et al.* (1994) was used. A characterization of the simulated systems is given in

Table 2. *Characterization of the simulated systems: the reported energies are an average over the last 60 ps of simulation*

	number of atoms	temperature K	initial box nm³	length of simulation ns	protein–solvent energy van der Waals (10^3 kJ mol^{-1})	electrostatic (10^3 kJ mol^{-1})
chloroform	9935	300	261	1.02	-1.73	-0.57
methanol	11362	300	261	1.02	-0.91	-3.47
water	19966	300	200	0.30	-0.64	-3.89
truncated SP-C in water at:						
300 K	5624	300	60[a]	1.03	-0.53	-2.97
500 K	11462	500	116	0.90	-0.45	-2.46

[a] A rectangular box was used instead of the truncated octahedron used in the other simulations.

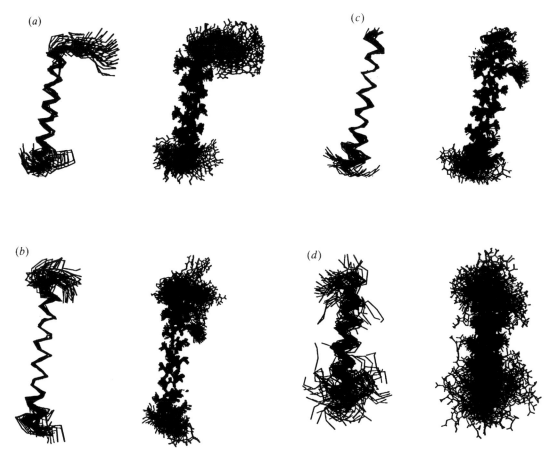

Figure 5. Superposition of averaged coordinates from 15 ps consecutive intervals from the 1 ns trajectories of the surfactant protein SP–C. Backbone C$_\alpha$ atoms (left) and all atoms (right) are displayed. (*a*) SP-C in chloroform, 300 K; (*b*) SP-C in methanol, 300 K; (*c*) truncated SP-C in water, 300 K; (*d*) truncated SP-C in water, 500 K. Only the C$_\alpha$ atoms of residues 9 to 28 of the complete peptide, LRIPCCPVNLKRLLVVVVVVVLVVVVIVGALLMGL, were included in the superposition. The molecules are depicted with the N-terminus up and C-terminus down. In truncated SP-C the first five residues are absent.

table 2. Owing to the small size of a water molecule, the simulation of SP-C in water is very expensive. Therefore, we decided to omit the first five N-terminal residues of SP-C, which were experimentally shown to be disordered, and simulated this truncated SP-C in water for a full nanosecond. A simulation of truncated SP-C in water at 500 K served as a control experiment and allowed a comparison with earlier high temperature unfolding studies of other proteins. The starting structure contained an extended chain (residues 1–8 and 35) and a helical part (residues 9–34). Further computational details are given by Kovacs *et al.* (1995).

Figure 5 shows a superposition of SP-C structures obtained from the four 1 ns MD simulations. At 300 K the α-helical conformation of SP-C is stable in all three solvents. It is most stable in water with a root mean square (r.m.s.) positional fluctuation of the C$_\alpha$ atoms of 0.069 nm, and of all atoms of 0.141 nm (residues 9–28). SP-C was least stable in chloroform with r.m.s. positional fluctuations of 0.126 nm for the C$_\alpha$ atoms and 0.290 nm for all atoms.

The simulations show that contrary to what might have been expected based on previously published values for the helix propensity of valines, the α-helical fold of the valyl-rich, predominantly hydrophobic

Figure 6. Residue–residue matrices showing the change in C_α–C_α atom distances for the DNA binding domain of the 434 repressor occurring in six MD simulations of 250 ps length starting from an equilibrated structure of the protein in aqueous solution. The symbols indicating the size of the changes are defined in the legend of figure 2. (*a*, *b*) Standard solvent interaction at (*a*) 300 K and (*b*) 350 K. (*c*, *d*) Modified Coulomb protein–solvent interaction: (*c*) 10% increased (SPM) and (*d*) 10% decreased (SPL). (*e*, *f*) Modified Coulomb solvent–solvent interaction: (*e*) 10% increased (SSM) and (*f*) 10% decreased (SSL).

peptide SP-C is remarkably stable in water, methanol and to a lesser extent in chloroform. This stability is primarily due to the additive effect of van der Waals interactions caused by the close packing of the branched aliphatic side-chains. This also prevents the solvent molecules from interacting with the protein backbone, especially in the poly-valyl part of Val-15–Val21. In the more polar solvents the interaction between hydrophobic side-chains is enhanced, producing increased helix stability. Even at elevated temperatures the poly-valyl stretch in the middle of the helical part of SP-C does not unfold. This does not mean that the α-helical fold of SP-C corresponds to the free-energy minimum in water. This cannot be determined from the current simulation. The results do demonstrate that the helix propensity of a given amino acid, in particular that of valine, can be highly sequence and environment specific.

5. STRUCTURAL STABILITY OF THE DNA BINDING DOMAIN OF THE 434 REPRESSOR IN DENATURING SOLVENTS

The three-dimensional structure of the 63-residue DNA binding domain of the 434 repressor in aqueous solution has been determined by NMR spectroscopy (Neri *et al.* 1992*a*). It consists of five α-helices and is not stabilized by disulphide bonds or metal ions. Yet, the folded conformation of this molecule is very stable, and even in 7 M urea a residual hydrophobic cluster is observed for residues 53–60 by NMR (Neri *et al.* 1992*b*).

In view of this experimental information, we decided to study the denaturation of the DNA-binding domain of the 434 repressor, under the influence of a denaturing solvent using non-equilibrium MD simulations. A simulation of a urea–water mixture would take a long time to equilibrate and sample the spatial distribution of the molecules properly. In short simulations, salt molecules would not have sufficient time to diffuse and equilibrate around the protein. A way around this limitation is to simulate a homogeneous aqueous environment but mimic the effect that the denaturant has on the protein conformation by changing the water–water or the water–protein interaction. First, the protein, immersed in a periodic truncated octahedron containing about 2000 water molecules, was equilibrated for 380 ps at 300 K and 1 bar. The corrected GROMOS force field and the SPC/E water model were used, and the equilibrated system was used to branch off five different simulations. In four of these the Coulomb interactions were modified as follows: the solvent–protein interaction was increased (SPM) or decreased (SPL) by 10%, or the solvent–solvent interaction was increased (SSM) or decreased (SSL) by 10%. For comparison, the equilibrium simulation was continued at 300 K, and a control simulation at 350 K was also performed: computational details are given in (Schiffer *et al.* 1995).

Figure 6 shows a comparison of the common starting structure with the structure after 250 ps of MD simulation of the six different systems. In the natural solvent (figure 6*a*) the protein maintains its native structure, apart from the C-terminal residue and a slight shift of the C-terminal ends of helices I and III with respect to each other. Of the four simulations with modified solvent interactions (*c–f*), the one in which the protein–solvent interaction was increased by 10% (SPM) shows by far the largest extent of protein denaturation. The denaturation is much larger than that obtained by raising the temperature to 350 K (figure 6*b*). This picture is confirmed if the simulations are extended by another 250 ps. The SPM denaturation process distorts the whole molecule, except for the five helices and the relative position of helix I with respect to helix V (figure 6*c*). Experimentally, the latter part of the molecule is seen to maintain residual structure in a high salt solution.

The simulations show that the protein is most easily denatured by increasing the protein–solvent interaction.

6. CONCLUSIONS

Using the proteins BPTI, HEW lysozyme, surfactant protein C and the DNA-binding domain of the 434 repressor as examples, it has been illustrated how an atomic picture of the onset of protein denaturation can be obtained by MD computer simulation. Denaturation can be induced by different driving forces, such as an increase of the temperature or pressure, a change of protein or solvent composition, or a modification of particular interactions in the molecular system. In all cases the process of protein denaturation is governed by the interplay of protein–protein, protein–solvent and solvent–solvent interactions. By monitoring the unfolding of a protein, the relative stability of the different parts of the protein can be determined and an indication of the responsible atomic interactions can be obtained.

When simulating non-equilibrium processes, there is generally little or no experimental information available with which the simulated results can be compared to establish the reliability of the latter. In addition, a direct comparison with results of hydrogen exchange labelling and other related experiments is made difficult by the large difference between simulated and experimental timescales, and by the possibility that unfolding and folding follow different pathways. This makes an unambiguous assessment of unfolding simulations difficult. Conversely, such simulations are useful because they give access to atomic details which are inaccessible by experiment. In cases where several models for a process have been proposed, the compatibility with the simulated results can be taken as supportive evidence in favour of a particular model.

REFERENCES

Berendsen, H.J.C., Grigera, J.R. & Straatsma, T.P. 1987 The missing term in effective pair potentials. *J. phys. Chem.* **91**, 6269–6271.

Berendsen, H.J.C., Postma, J.P.M., van Gunsteren, W.F. & Hermans, J. 1981 Interaction models for water in relation to protein hydration. In *Intermolecular forces* (ed. B. Pullman), pp. 331–342. Dordrecht: Reidel.

Brunne, R.M. & van Gunsteren, W.F. 1993 Dynamical properties of bovine pancreatic trypsin inhibitor from a molecular dynamics simulation at 5000 atm. *FEBS Lett.* **323**, 215–217.

Caflisch, A. & Karplus, M. 1994 Molecular dynamics simulation of protein denaturation: Solvation of the hydrophobic cores and secondary structure of barnase. *Proc. natn. Acad. Sci. U.S.A.* **91**, 1746–1750.

Covell, D.G. & Jernigan, R.L. 1990 Conformations of folded proteins in restricted spaces. *Biochemistry* **29**, 3287–3294.

Daggett, V. 1993 A model for the molten globule state of CTF generated using molecular dynamics. In *Techniques in protein chemistry IV* (ed. R.H. Angeletti), pp. 525–532.

Daggett, V. & Levitt, M. 1992 A model of the molten globule state from molecular dynamics simulations. *Proc. natn. Acac. Sci. U.S.A.* **89**, 5142–5146.

Daggett, V. & Levitt, M. 1993 Protein unfolding pathways explored through molecular dynamics simulations. *J. molec. Biol.* **232**, 600–619.

Darby, N.J., van Mierlo, C.P.M., Scott, G.H.E., Neuhaus, D. & Creighton, T.E. 1992 Kinetic roles and conformational properties of the non–native two-disulphide intermediates in the refolding of bovine pancreatic trypsin inhibitor. *J. molec. Biol.* **22**, 905–911.

Dietz, W. & Heinzinger, K. 1984 Structure of liquid chloroform. A comparison between computer simulation and neutron scattering results. *Ber. BunsenGes. phys. Chem.* **88**, 543–546.

Dietz, W. & Heinzinger, K. 1985 A molecular dynamics study of liquid chloroform. *Ber. BunsenGes. phys. Chem.* **89**, 968–977.

Dill, K.A. 1985 Theory for the folding and stability of globular proteins. *Biochemistry* **24**, 1501–1509.

Dill, K.A. 1990 Dominant forces in protein folding. *Biochemistry* **29**, 7133–7155.

Dobson, C.M., Evans, P.A. & Radford, S.E. 1994 Understanding how proteins fold: the lysozyme story so far. *TIBS* **19**, 31–37.

Goldenberg, D.P. 1992 Native and non-native intermediates in the BPTI folding pathway. *TIBS* **17**, 257–261.

Hao, M.-H., Pincus, M.R., Rackovsky, S. & Scheraga, H.A. 1993 Unfolding and refolding of the native structure of bovine pancreatic trypsin inhibitor studied by computer simulation. *Biochemistry* **32**, 9614–9631.

Hünenberger, P.H., Mark, A.E. & van Gunsteren, W.F. 1995 Computational approaches to study protein unfolding: Hen egg white lysozyme as a case study. *Proteins.* (In the press.)

Johansson, J., Curstedt, T. & Robertson, B. 1994a The proteins of the surfactant system. *Eur. Resp. J.* **7**, 372–391.

Johansson, J., Szyperski, T., Curstedt, T. & Wüthrich, K. 1994b The NMR structure of the pulmonary surfactant-associated polypeptide sp–c in an apolar solvent contains a valyl-rich α-helix. *Biochemistry* **33**, 6015–6023.

Jones, D.T., Taylor, W.R. & Thornton, J.M. 1992 A new approach to protein fold recognition. *Nature, Lond.* **358**, 86–89.

Kabsch, W. & Sander, C. 1983 Dictionary of protein secondary structure: Pattern recognition of hydrogen-bonded and geometrical features. *Biopolymers* **22**, 2577–2637.

Kitchen, D.B., Reed, L.H. & Levy, R.M. 1992 Molecular dynamics simulation of solvated protein at high pressure. *Biochemistry* **31**, 10083–10093.

Kocher, J.-P., Rooman, M.J. & Wodak, S.J. 1994 Factors influencing the ability of knowledge-based potentials to identify native sequence-structure matches. *J. molec. Biol.* **235**, 1598–1613.

Kolinski, A. & Skolnick, J. 1994 Monte Carlo simulations of protein folding. I. Lattice model and interaction scheme. *Proteins* **18**, 338–352.

Kovacs, H., Mark, A.E., Johansson, J. & van Gunsteren, W.F. 1995 The effect of environment on the stability of an integral membrane helix: Molecular dynamics simulations of surfactant protein C in chloroform, methanol and water. *J. molec. Biol.* (In the press.)

Kundrot, C.E. & Richards, F.M. 1987 Crystal structure of hen egg-white lysozyme at a hydrostatic pressure of 1000 atmospheres. *J. molec. Biol.* **193**, 157–170.

Levitt, M. & Warshel, A. 1975 Computer simulation of protein folding. *Nature, Lond.* **253**, 694–698.

Lide, D.R. 1993 *Handbook of chemistry and physics.* Boca Raton, Florida: CRC Press.

Lüthy, R., Bowie, J.U. & Eisenberg, D. 1992 Assessment of protein models with three-dimensional profiles. *Nature, Lond.* **356**, 83–85.

Maiorov, V.N. & Crippen, G.M. 1992 Contact potential that recognizes the correct folding of globular proteins. *J. molec. Biol.* **227**, 876–888.

Mark, A.E. & van Gunsteren, W.F. 1992 Simulation of the thermal denaturation of hen egg white lysozyme: Trapping the molten globule state. *Biochemistry* **31**, 7745–7748.

Mark, A.E., van Helden, S.P., Smith, P.E., Janssen, L.H.M. & van Gunsteren, W.F. 1994 Convergence properties of free energy calculations: α-cyclodextrin complexes as a case study. *J. Am. chem. Soc.* **116**, 6293–6302.

Neri, D., Billeter, M. & Wüthrich, K. 1992a Determination of the nuclear magnetic resonance solution structure of the DNA-binding domain (residues 1 to 69) of the 434 repressor and comparison with the X-ray crystal structure. *J. molec. Biol.* **223**, 743–767.

Neri, D., Billeter, M., Wider, G. & Wüthrich, K. 1992b NMR determination of residual structure in a urea-denatured protein, the 434-repressor. *Science, Wash.* **257**, 1559–1563.

Novotny, J., Bruccoleri, R. & Karplus, M. 1984 An analysis of incorrectly folded protein models. Implication for structure prediction. *J. molec. Biol.* **177**, 787–818.

Ouzounis, C., Sander, C., Scharf, M. & Schneider, R. 1993 Prediction of protein structure by evaluation of sequence–structure fitness. Aligning sequences to contact profiles derived from three-dimensional structures. *J. molec. Biol.* **232**, 805–825.

Sali, A., Shakhnovich, E. & Karplus, M. 1994 Kinetics of protein folding. A lattice model study of the requirements for folding to the native state. *J. molec. Biol.* **235**, 1614–1636.

Samarasinghe, S.D., Campbell, D.M., Jonas, A. & Jonas, J. 1992 High-resolution NMR study of the pressure-induced unfolding of lysozyme. *Biochemistry* **31**, 7773–7778.

Schiffer, C.A. & van Gunsteren, W.F. 1995 Structural stability of disulfide mutants of BPTI: A molecular dynamics study. (Submitted.)

Schiffer, C.A., Dötsch, V. & van Gunsteren, W.F. 1995 Exploring the role of solvent in the denaturation of a protein: A molecular dynamics study on the DNA binding domain of the 434 repressor. *Biochemistry.* (Submitted.)

Seetharamulu, P. & Crippen, G.M. 1991 A potential function for protein folding. *J. math. Chem.* **6**, 91–110.

Shi Yun-yu, Mark, A.E., Wang Cun-xin, Huang Fuhua, Berendsen, H.J.C. & van Gunsteren, W.F. 1993 Can the stability of protein mutants be predicted by free energy calculations? *Protein Engng.* **6**, 289–295.

Stouten, P.F.W. 1989 Small molecules in the liquid, solid and computational phases. Dissertation, University of Utrecht, The Netherlands.

Tirado-Rives, J. & Jorgensen, W.L. 1993 Molecular dynamics simulations of the unfolding of apomyoglobin in water. *Biochemistry* **32**, 4175–4184.

Unger, R. & Moult, J. 1993 Genetic algorithms for protein folding simulations. *J. molec. Biol.* **231**, 75–81.

van Gunsteren, W.F. & Berendsen, H.J.C. 1987 *Groningen molecular simulation* (GROMOS) library manual. Groningen, The Netherlands: Biomos.

van Mierlo, C.P.M., Darby, N.J. & Creighton, T.E. 1992 The partially folded conformation of the Cys-30 Cys-51 intermediate in the disulfide folding pathway of bovine pancreatic trypsin inhibitor. *Proc. natn. Acad. Sci. U.S.A.* **89**, 6775–6779.

Weissman, J.S. & Kim, P.S. 1991 Reexamination of the folding of BPTI: Predominance of native intermediates. *Science, Wash.* **253**, 1386–1393.

Models of cooperativity in protein folding

HUE SUN CHAN, SARINA BROMBERG AND KEN A. DILL

Department of Pharmaceutical Chemistry, Box 1204, University of California, San Francisco, California 94143–1204, U.S.A.

SUMMARY

What is the basis for the two-state cooperativity of protein folding? Since the 1950s, three main models have been put forward.

1. In 'helix-coil' theory, cooperativity is due to local interactions among near neighbours in the sequence. Helix-coil cooperativity is probably not the principal basis for the folding of globular proteins because it is not two-state, the forces are weak, it does not account for sheet proteins, and there is no evidence that helix formation precedes the formation of a hydrophobic core in the folding pathways.

2. In the 'sidechain packing' model, cooperativity is attributed to the jigsaw-puzzle-like complementary fits of sidechains. This too is probably not the basis of folding cooperativity because exact models and experiments on homopolymers with sidechains give no evidence that sidechain freezing is two-state, sidechain complementarities in proteins are only weak trends, and the molten globule model predicted by this model is far more native-like than experiments indicate.

3. In the 'hydrophobic core collapse' model, cooperativity is due to the assembly of non-polar residues into a good core. Exact model studies show that this model gives two-state behaviour for some sequences of hydrophobic and polar monomers. It is based on strong forces. There is considerable experimental evidence for the kinetics this model predicts: the development of hydrophobic clusters and cores is concurrent with secondary structure formation. It predicts compact denatured states with sizes and degrees of disorder that are in reasonable agreement with experiments.

1. INTRODUCTION

What makes protein folding cooperative? Protein folding, at least for small single-domain proteins, is known to involve two-state transitions (Lumry *et al.* 1966; Privalov 1979). The purpose of this review is to describe one-state and two-state transitions, to review various models for the physical basis of the two-state protein folding process, and to describe how the nature of folding kinetics and denatured states must follow directly from the physical basis for the cooperativity. We review three models for cooperativity: the helix-coil model, the sidechain packing model, and the hydrophobic collapse model. We conclude that protein folding cooperativity is modelled most simply as a heteropolymer hydrophobic collapse process.

What is cooperativity? Figure 1*a* shows a non-cooperative process in which a system gradually changes from state A to B as a function of temperature or denaturant. Figure 1*b, c* shows the experimental signature of a cooperative process, namely sigmoidal behaviour. Cooperativity may be of two different types: two-state (first order) or one-state (higher order). Neither the observation of a sigmoidal transition curve, nor the observation of a peak of heat absorption by scanning calorimetry can distinguish a two-state from a one-state cooperative transition. The definitive way to distinguish experimentally between these two types of transition is by an analysis of the populations. At the midpoint of a two-state transition,

two identifiable states will be populated. In contrast, one-state behaviour is defined by a single broad population peak at the midpoint. It follows that two-state cooperativity implies a free energy surface with two minima separated by a barrier. One-state cooperativity implies a single free energy minimum. Whether the cooperativity of a process is one- or two-state does not necessarily depend on the steepness of its sigmoidal curve. Figure 2 shows examples from exact model studies indicating that the steepness of a sigmoidal curve in a one-state process can be just as great as in a two-state process. Cooperativity depends, rather, on a molecular property that is more difficult to measure: the distribution of the underlying populations.

The discrimination of one-state from two-state cooperativity determines fundamental information about the free energy surface, and the underlying molecular mechanism of the cooperative process. How is two-state behaviour determined experimentally? There are various methods.

1. As noted above, the most definitive determination of two-state behaviour is to observe two distinct populations near the midpoint of the transition. For slowly exchanging systems, this can be done by transport methods (Cann 1970) such as size exclusion chromatography (Uversky 1993) or gel electrophoresis (Creighton 1986). Sometimes a spectroscopic method will give distinct signatures for the native molecule in native conditions and the highly unfolded molecule in

Phil. Trans. R. Soc. Lond. B (1995) **348**, 61–70

Printed in Great Britain

61

© 1995 The Royal Society

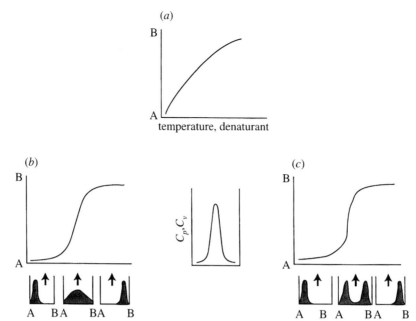

Figure 1. What is cooperativity? Two states, A and B, such as native and denatured states, can change populations with temperature, denaturant, pH, salt, etc. (*a*) Gradual change, no cooperativity. (*b*) Cooperative transition of the one-state type. (*c*) Cooperative transition of the two-state type. Both one-state and two-state transitions can have sigmoidal behaviour and heat absorption (a peak in the C_p or C_v plot); they cannot be distinguished on these bases, or from the steepness of the sigmoidal curve. The main distinction is whether there is one broad peak involving high populations of 'intermediates' near the denaturant midpoint (one-state) or whether there are two populated states and less intermediate population (small plots at the bottom of (*b*) and (*c*)).

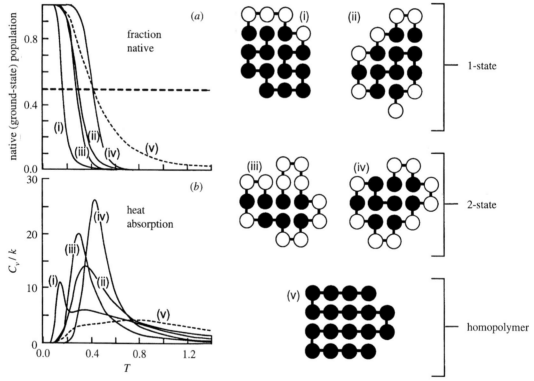

Figure 2. Cooperativity among HP lattice model sequences (H, hydrophobic, black beads; P, polar, white beads). All five of these sequences collapse cooperatively, i.e. have sigmoidal changes in the native population with temperature T, and have heat absorption (C_v) peaks. Sequence (v) is a homopolymer shown in one of its 1673 maximally compact states; all four other sequences are shown in their unique native states. Only an analysis of the underlying populations can show that the cooperativity for sequences (i), (ii), and (v) is one-state, and for sequences (iii) and (iv) is two-state. Figure 3 demonstrates this by showing the underlying populations for sequences (i) and (iv).

strongly denaturing conditions. Spectroscopic signal changes through the denaturation midpoint that can be modelled as a linear combination of the signals from the two limiting states are an indication of two-state behaviour. On the other hand, if the spectroscopic signal near the denaturation midpoint cannot be fitted

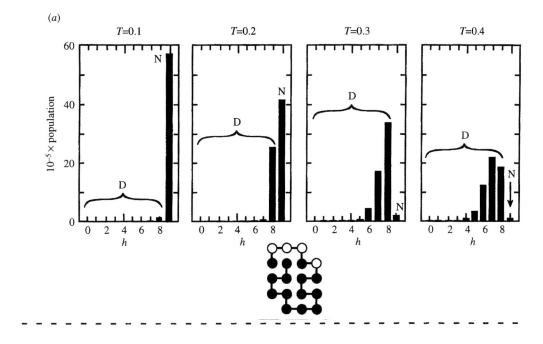

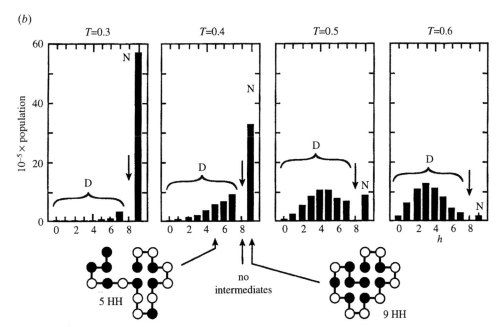

Figure 3. Population analysis of conformations for the HP lattice model sequences in figure 2. (*a*) Sequence (i) shows one-state behaviour. (*b*) Sequence (iv) shows two-state behaviour. On the horizontal axis, *h* is the number of HH contacts (*h* = 9 is native for both sequences); the vertical axis shows the populations. An example denatured conformation with *h* = 5 is shown in (*b*). At low temperature most molecules are in their native states. At higher temperature most molecules are denatured. Near the denaturation midpoint (*a*) shows a high population of intermediates, while (*b*) shows no population of *h* = 8 intermediate conformations. Notice also that the denatured state shifts population with temperature: at low temperatures the denatured molecules are very compact (high *h*), whereas at higher temperatures the denatured molecules are more unfolded.

as a simple linear combination of the two limiting state signals, it is evidence for one-state behaviour. Now a significant population of 'intermediate' molecules, which may have different spectral signatures, may contribute to the total signal.

2. Another measure of two-state behaviour is the ratio of the calorimetric enthalpy (a model-independent quantity, determined directly from experiments), to the van't Hoff enthalpy (a model-dependent quantity based on certain thermodynamic assumptions) (Privalov 1979; Privalov & Gill 1988).

3. A less definitive measure of two-state behaviour is the observation that different experimental measures show coincident sigmoidal curves.

The reason that (3) (and possibly (2)) are less definitive measures of two-state behaviour is that both are based on the assumption, not always correct, that the two states of proteins are as fixed and simple as the two different states observed in many small-molecule processes. However, the second state of proteins, the denatured state, is more complex. The radius and other properties of the denatured molecules can change

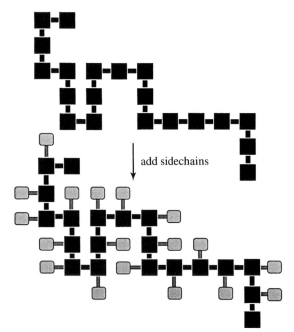

Figure 4. Simple exact side-chain model. Taking the linear chain lattice model to represent the main chain (upper figure), a side-chain model is created by attaching a single side chain unit to each main chain monomer. To represent side chain rotameric degrees of freedom, each side chain unit has the freedom to occupy any one empty lattice site adjacent to its corresponding main chain monomer (lower figure); see Bromberg & Dill (1994).

with external conditions. Two-state protein unfolding behaviour can lead to situations in which different experimental measures will not have coincident sigmoidal curves (see appendix, Dill & Shortle 1991). Coincident curves give evidence supporting two-state behaviour, but non-coincident curves are not definitive evidence against two-state behaviour.

What is the physical basis for cooperativity in protein folding? The first model for cooperativity in biopolymer conformational transitions is the helix-coil theory (Schellman 1958; Zimm & Bragg 1959; Lifson & Roig 1961; Poland & Scheraga 1970). This model represents the cooperativity that comes from the influences of nearest neighbours along the chain. A residue has more propensity to be in a helical configuration if its neighbour in the sequence is helical than if its neighbour is in a coil configuration. Helix-coil theories are based on the one-dimensional Ising model, for which the cooperativity has been proven to be one-state (Stanley 1971). Thus local interactions along the chain are not sufficient to account for the two-state cooperativity of protein folding.

Another possibility is that protein cooperativity originates in the non-local interactions that cause the sharp collapse of hydrophobic polymers in water (Fujishige *et al.* 1989; Ricka *et al.* 1990; Meewes *et al.* 1991; Tiktopulo *et al.* 1994). Ptitsyn and coworkers were the first to develop a model showing that homopolymer collapse processes would be cooperative (Ptitsyn *et al.* 1968). However, many theoretical model studies beginning in the 1970s (de Gennes 1975; Moore 1977; Post & Zimm 1979; Sanchez 1979; Grosberg & Khokhlov 1987) determined that homopolymer col-

lapse is only one-state, unless the chain stiffness is high. Thus the collapse of flexible homopolymers could not account for the two-state cooperativity of protein folding. Recent experiments show that the collapse of homopolymers of poly-N(isopropylacrylamide) (PNIPAM) is likely to be a one-state process (Tiktopulo *et al.* 1994).

As homopolymer collapse and helix-coil models could not account for the two-state cooperativity of globular proteins, two different models for protein folding cooperativity were developed during the 1980s. A mean-field theory was developed, based on the assumption that protein (heteropolymer) collapse differs from homopolymer collapse in that some sequences of hydrophobic and polar monomers can cooperatively form good hydrophobic cores (Dill 1985; reviewed in Chan *et al.* 1992; Dill & Stigter 1994). We call this the hydrophobic core (HC) model. On the other hand, Shakhnovich and Finkelstein (1989) regarded cooperativity as arising from jigsaw-puzzle-like packing and interdigitation among sidechains. They modelled unfolding cooperativity arising as sidechains suddenly become freed from their local packing constraints at a critical disjuncture point in the chain expansion. We call this the sidechain packing (SP) model. The HC model assumed two-state behaviour, rather than proving it, and was based on two mean-field approximations. The SP model was based on many different assumptions and approximations. It was not clear whether the two-state cooperativity assumed and implied in those models could be derived on more rigorous grounds. To resolve the basis for cooperativity, we have recently resorted to simplified models in which the partition function can be known exactly and without approximation. This review summarizes those results.

2. HYDROPHOBIC CORE MODEL

By exact enumeration, we have explored the full conformational space of some HP lattice model chains on two-dimensional square lattices (Lau & Dill 1989, 1990; Chan & Dill 1991; Chan *et al.* 1992). Figure 2 shows the native population at various temperatures, for each of four different unique sequences, and a homopolymer. Each unique sequence folds only to a single lowest energy native state, whereas the ground ('native') state of the homopolymer is the full ensemble of maximally compact conformations (Chan & Dill 1989). Figure 2 shows that all these sequences have a sigmoidal variation of the fraction of molecules that are native versus temperature, and that all these sequences have peaks of heat absorption upon folding; observations generally taken to signal cooperative behaviour. But these simple tests (sigmoidal behaviour, heat absorption) alone do not tell us whether the transitions are one-state or two-state. Such information is revealed most definitively by a full population analysis; see figure 3. Exact calculations show rigorously that two-state behaviour is observed in two of the sequences, and one-state behaviour is found for the other two unique sequences. Thus it is clear that the collapse of heteropolymers and the formation of a good hydrophobic core is a sufficient physical basis to obtain the type of two-state behaviour observed in the folding of globular proteins. But within the HP model, not all sequences will do this. For example, the homopolymer sequence HHHH . . . . H collapses with a one-state transition.

3. SIDECHAIN PACKING MODEL

Figure 4 shows the simple exact model we have used to study sidechain packing (Bromberg & Dill 1994) by exhaustive enumeration for short chains, and by Monte Carlo sampling for longer chains. In this simple model, sidechains are represented as monomers of uniform size and shape.

Figure 5*b* shows the computed conformational entropies due to sidechain degrees of freedom and excluded volume. As longer chains expand in denaturation from the native state, sidechain freedom and conformational entropy increase rapidly at first, then more slowly while the volume explored by the backbone continues to expand. The sharp gain in freedom near the native state is like an 'unfreezing' of the sidechains. In contrast, figure 5*a* shows the model of Shakhnovich & Finkelstein (1989), which assumes a 'critical disjuncture point' at which the sidechains suddenly become unlocked, not at the earliest stages of chain expansion, but at a critical point of expansion, around a 25% volume increase. This disjuncture point

is the basis for the cooperativity in their model, but such a critical disjuncture point is not found in the simple exact model study.

In light of the fact that neither the model of Shakhnovich & Finkelstein, nor our simple exact sidechain model accurately represents real protein sidechains, what conclusions can be drawn? We believe the exact sidechain model, in its simplicity, bears closer resemblance to real proteins, insofar as sidechains do not need to be precisely matched in shape and locked together for a protein to achieve either (i) tight packing, or (ii) native topology. The following data support this view.

Sidechains can achieve native packing densities without locking together according to a precise stereochemical code:

1. In the crystal structures of the Protein Data Bank (PDB) (Bernstein *et al.* 1977; Abola *et al.* 1987), sidechains show little preference for particular partners to mutually bury surface area (Behe *et al.* 1991).

2. Pairs of hydrophobic core sidechains in the PDB show little preference for particular mutual orientations (Singh & Thornton 1990, 1992).

3. Native packing density in computer simulated protein structures is readily achieved even by misfolded sequences (Novotny *et al.* 1984).

4. Protein interiors are dynamic: small molecules rapidly diffuse to buried residues (Gurd & Rothgeb 1979; Karplus & McCammon 1981); and aromatic

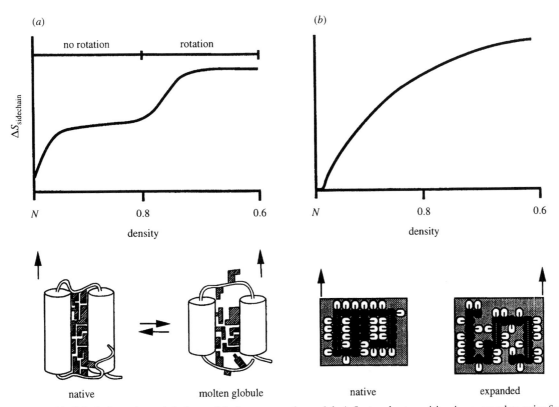

Figure 5. (*a*) Sidechain molten globule model: jig saw puzzle model. A first-order transition is proposed to arise from the native to molten globule state because of a sharp increase in sidechain rotational entropy at a critical disjuncture point. Backbone and secondary structures are assumed fixed in native-like conformations, and thus assumed to be independent of sidechain freedom (Shakhnovich & Finkelstein 1989; Ptitsyn 1992). (*b*) Nuts and bolts model. By exact enumeration in a simplified model of sidechains, the sidechain rotational entropy is shown to increase most sharply even at the earliest expansions from the native state, implying no critical disjuncture point (Bromberg & Dill 1994). Sidechains and backbone are found to be strongly coupled. It is proposed that a critical disjuncture point, corresponding to sidechain unlocking, is not the defining characteristic of compact denatured states.

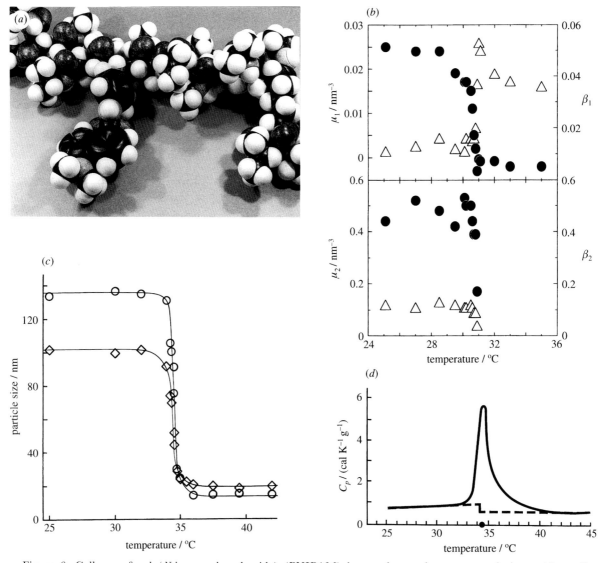

Figure 6. Collapse of poly(*N*-isopropylacrylamide) (PNIPAM) homopolymers in aqueous solutions with small amount of surfactant to suppress aggregration. (*a*) Space-filling model of a section of the polymer. (*b*) Homopolymer collapse freezes out the motions of the sidechains. Time-resolved measurements of fluorescence polarization anisotropy are used to monitor sidechain motions due to the temperature-induced conformational transition. The plots show reduced re-orientational relaxation rates μ_i and their amplitudes β_i versus temperature: (left, filled circles) μ_1, (right, open triangles) β_1, reflecting backbone motions; (left, filled circles) μ_2, (right, open triangles) β_2, reflecting mainly local sidechain motions. Both types of motions undergo a 'freezing' transition at the same temperature around 31 °C. The chains have approximately 3100 monomer units. (*a*) and (*b*) are reproduced from Binkert *et al.* (1991). (*c*) Hydrodynamic radius (open diamonds) and radius of gyration (open circles) of PNIPAM as a function of temperature (data from Meewes *et al.* 1991). (*d*) Temperature dependence of the partial specific heat capacity of PNIPAM. Dashed lines show the extrapolations of heat capacity curves from low and high temperatures to the middle of the transition region. Reproduced from Tiktopulo *et al.* (1994). The number of monomers per chain in experiments (*c*) and (*d*) are approximately 62 000.

rings rotate with low activation energy barriers (Wuthrich & Wagner 1978). The temperature factors of protein crystals (Artymiuk *et al.* 1979; Frauenfelder *et al.* 1979) show that the backbone atoms of an amino acid residue in a protein core are more rigidly immobilized than its side-chain atoms.

5. Cavities created by mutations in protein cores are not rigid like jigsaw puzzles lacking a piece. (McRee *et al.* 1990; Eriksson *et al.* 1992; Varadarajan & Richards 1992). Main chain readjustments accomodate many mutations that change side chain volumes (Baldwin *et al.* 1993).

6. The kinetic bottleneck to protein folding is likely to be the breaking of non-native bonds or contacts, rather than side chain locking. It has been shown that a single non-native heme interaction slows the folding of cytochrome *c* by orders of magnitude (Sosnick *et al.* 1994), and non-native disulphide bonds or aromatic interactions occur in folding intermediates of hen lysozyme (Chaffotte *et al.* 1992; Hooke *et al.* 1994). Other examples of non-native contacts are reviewed by Creighton (1994).

Proteins can achieve native topologies without unique sidechain packing:

1. Topologically similar proteins can have differently packed cores (Swindells & Thornton 1993).

2. Proteins can maintain native-like topology in states that lack native tight packing (Feng *et al.* 1994; Hughson *et al.* 1994; Peng & Kim 1994).

3. The fold of some proteins, such as globins, can be achieved by sequences that are less than 20% identical (Bashford *et al.* 1987).

4. Proteins show considerable structural tolerance for mutations that change core side chain size and shape (Matthews 1987, 1993; Lim & Sauer 1991; Keefe *et al.* 1993, Richards & Lim 1994).

Like proteins (Richards 1974, 1977; Richards & Lim 1994), the simple exact sidechain model fills space without large voids. The configurational freedom of the sidechains is linked to the freedom of the backbone. These model chains collapse to configurations with distributions of contacts among and between the main chain and the sidechains similar to the distributions found in the proteins of the PDB (Bromberg & Dill 1994). Nevertheless, because this model does not treat sidechains of different sizes or shapes, we may rigorously conclude only that this model, by itself, provides no basis for two-state cooperativity.

Experiments confirm this conclusion. Experiments with a protein-model polymer PNIPAM indicate that sidechain packing does not cause two-state behaviour. PNIPAM collapses with a very sharp sigmoidal cooperativity, and has an associated peak in heat absorption; see figure 6. The very steepness of the transition (the full collapse process takes place over only about 1 °C) is not a basis for concluding the transition is two-state. Experiments of Tiktopulo *et al.*(1994) show that the ratio of calorimetric to van't Hoff enthalpies equals 120, which is strong evidence that this collapse process is, as predicted by homopolymer theories, only one-state.

Experiments by Binkert *et al.*(1991) show that the rates of sidechain motions measured by fluorescence anisotropy slow dramatically at the transition (see figure 6). Taken together, these experiments indicate that molecules can have sidechains that 'freeze' upon collapse, as they do in proteins and in the exact sidechain model, without causing the transition to be two-state. These experiments and the simple exact model imply that sidechain freezing does occur in polymer collapse, but that it need not cause two-state behaviour. These are only experimental and theoretical models, not real proteins, but they give no evidence that more realistic models would show that sidechain freezing is the principal cause of two-state cooperativity in proteins.

4. IMPLICATIONS FOR PROTEIN FOLDING KINETICS AND COMPACT DENATURED STATES

How we view the forces causing cooperativity determines how we understand the folding kinetics and the denatured states of proteins. For example, the SP model (Shakhnovich & Finkelstein 1989; Ptitsyn 1987; Karplus & Shakhnovich 1992) predicts that the native state goes to a denatured state that maintains the secondary structures of the native molecule, with only

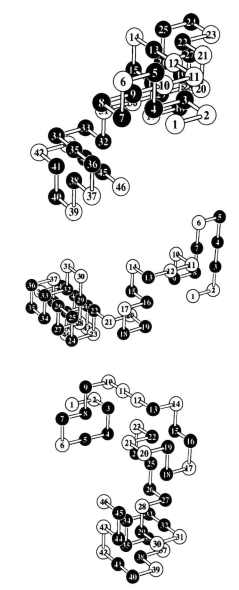

Figure 7. The hydrophobic core model implies that compact denatured states are sometimes broad ensembles of different backbone configurations, depending on sequence and external conditions. These three-dimensional HP lattice model conformations represent three different chain configurations in the ensemble of a compact denatured state of a protein. (From Lattman *et al.* 1994.)

the sidechains freed. This was the basis for the 'molten globule' model of compact denatured states; see figure 5. On the other hand, the HC model transition goes from native to compact denatured states with a much broader ensemble of backbone and sidechain conformations, ensembles of secondary structures and hydrophobic clusters, radii slightly larger than native, and bimodal light scattering $P(r)$ curves (Lattman *et al.* 1994), all of which are strongly dependent on the amino acid sequence and external conditions; see figure 7.

Similarly, folding kinetics is dependent on the nature of cooperativity. Whereas models based on helix-coil-like cooperativity lead to kinetics in which isolated secondary structures are predicted to form as the earliest steps (Ptitsyn *et al.* 1972; Karplus & Weaver, 1976, 1994; Kim & Baldwin 1982; Baldwin 1989), hydrophobic core cooperativity models imply that folding is driven by the hydrophobic zipping and assembly process, which causes concurrent secondary structure formation (Dill *et al.* 1993; Fiebig & Dill 1993; Lattman *et al.* 1994). This is consistent with data showing concurrent collapse and some secondary structure formation in early stages of folding (Gilmanshin & Ptitsyn 1987; Semisotnov *et al.* 1987; Briggs & Roder 1992; Elöve *et al.* 1992; Serrano *et al.* 1992; Jennings & Wright 1993; Itzhaki *et al.* 1994; Nishii *et al.* 1994, reviewed in Barrick & Baldwin 1993).

5. CONCLUSIONS

We have discussed three possible models for the two-state cooperativity in the folding of globular proteins. The limitations of the helix-coil model of biopolymer cooperativity for the basis of globular protein folding are that: (i) helical propensities are weak in aqueous solution (Chakrabartty *et al.* 1991, 1994; Scholtz *et al.* 1991; Scholtz & Baldwin 1992); (ii) helical propensities cannot account for sheet proteins; (iii) helix-coil transitions give only one-state cooperativity; and (iv) evidence is mounting for helix formation concurrent with hydrophobic clustering in the first stages of folding kinetics (Dill *et al.* 1995). The limitations of the sidechain packing model are that: (i) sidechains do not have strong packing preferences; (ii) the transition is probably one-state; and (iii) the predicted molten globule is more rigid and native-like, and less dependent on external conditions than are indicated by experiments (reviewed in Dill & Shortle 1991; Shortle *et al.* 1992; Shortle 1993). We believe the simplest basis for understanding protein cooperativity is in terms of hydrophobic and polar sequences that can form good hydrophobic cores. The driving force (non-polar association in water) is strong, and the folding kinetics of collapse accompanied by secondary structure formation is consistent with experiments and the secondary structure length distribution of proteins in the PDB (Chan & Dill 1990).

We thank the National Institutes of Health for financial support.

REFERENCES

Abola, E.E., Bernstein, F.C., Bryant, S.H., Koetzle, T.F. & Weng, J. 1987 Protein data bank. In *Crystallographic databases – information content, software systems, scientific applications* (ed. F.H. Allen, G. Bergerhoff & R. Seivers), pp.107–132. Bonn, Cambridge and Chester: Data Commission of the International Union of Crystallography.

Artymiuk, P.J., Blake, C.C., Grace, D.E., Oatley, S.J., Phillips, D.C. & Sternberg, M.J. 1979 Crystallographic studies of the dynamic properties of lysozyme. *Nature, Lond.* **280**, 563–568.

Baldwin, R.L. 1989 How does protein folding get started? *Trends Biochem. Sci.* **14**, 291–294.

Baldwin, E.P., Hajiseyedjavadi, O., Baase, W.A. & Matthews, B.W. 1993 The role of backbone flexibility in the accomodations of variants that repack the core of T4 lysozyme. *Science, Wash.* **262**, 1715–1718.

Barrick, D. & Baldwin, R.L. 1993 Stein and Moore Award address. The molten globule intermediate of apomyoglobin and the process of protein folding. *Protein Sci.* **2**, 869–876.

Bashford, D., Chothia, C. & Lesk, A.M. 1987 Determinants of a protein fold. Unique features of the globin amino acid sequences. *J. molec. Biol.* **196**, 199–216.

Behe, M.J., Lattman, E.E. & Rose, G.D. 1991 The protein-folding problem: the native fold determines packing, but does packing determine the native fold? *Proc. natn. Acad. Sci. U.S.A.* **88**, 4195–4199.

Bernstein, F.C., Koetzle, T.F., Williams, G.J.B., Meyer, E.F. Jr, Brice, M.D., Rodgers, J.R., Kennard, O., Shimanouchi, T. & Tasumi, M. 1977 The Protein Data Bank: a computer-based archival file for macromolecular structures. *J. molec. Biol.* **112**, 535–542.

Binkert, Th., Oberreich, J., Meewes, M., Nyffenegger, R. & Ricka, J. 1991 Coil-globule trasition of poly(N-isopropylacrylamide): a study of segment mobility by fluorescence depolarization. *Macromolecules* **24**, 5806–5810.

Briggs, M.S. & Roder, H. 1992 Early hydrogen-bonding events in the folding reaction of ubiquitin. *Proc. natn. Acad. Sci. U.S.A.* **89**, 2017–2021.

Bromberg, S. & Dill, K.A. 1994 Sidechain entropy and packing in proteins. *Protein Sci.* **3**, 997–1009.

Cann, J.R. 1970 *Interacting macromolecules: the theory and practice of their electrophoresis, ultracentrifugation, and chromatography.* New York: Academic Press.

Chaffotte, A.F., Guillou, Y. & Goldberg, M.E. 1992 Kinetic resolution of peptide bond and side chain far-UV circular dichroism during the folding of hen egg white lysozyme. *Biochemistry* **31**, 9694–9702.

Chakrabartty, A., Schellman, J.A. & Baldwin, R.L. 1991 Large differences in the helix propensities of alanine and glycine. *Nature, Lond.* **351**, 586–588.

Chakrabartty, A., Kortemme, T. & Baldwin, R.L. 1994 Helix propensities of the amino acids measured in alanine-based peptides without helix-stabilizing side-chain interactions. *Protein Sci.* **3**, 843–852.

Chan, H.S. & Dill, K.A. 1989 Compact polymers. *Macromolecules* **22**, 4559–4573.

Chan, H.S. & Dill, K.A. 1990 Origins of structure in globular proteins. *Proc. natn. Acad. Sci. U.S.A.* **87**, 6388–6392.

Chan, H.S. & Dill, K.A. 1991 'Sequence space soup' of proteins and copolymers. *J. chem. Phys.* **95**, 3775–3787.

Chan, H.S., Dill, K.A. & Shortle, D. 1992 Statistical mechanics and protein folding. In *Princeton lectures on biophysics* (ed. W. Bialek), pp. 69–173. Singapore: World Scientific.

Creighton, T.E. 1986 Detection of folding intermediates using urea-gradient electrophoresis. *Meth. Enzymol.* **131**, 156–172.

Creighton, T.E. 1994 The energetic ups and downs of protein folding. *Nature Struct. Biol.* **1**, 135–138.

de Gennes, P.-G. 1975 Collapse of a polymer chain in poor solvents. *J. Phys. Lett., Paris* **36**, L55–L57.

Dill, K.A. 1985 Theory for the folding and stability of globular proteins. *Biochemistry* **24**, 1501–1509.

Dill, K.A. & Shortle, D. 1991 Denatured states of proteins. *A. Rev. Biochem.* **60**, 795–825.

Dill, K.A., Fiebig, K.M. & Chan, H.S. 1993 Cooperativity in protein folding kinetics. *Proc. natn. Acad. Sci. U.S.A.* **90**, 748–52.

Dill, K.A. & Stigter, D. 1994 Modeling protein stability as heteropolymer collapse. *Adv. Protein Chem.* (In the press.)

Dill, K.A., Bromberg, S., Yue, K., Fiebig, K.M., Yee, D.P., Thomas, P.D. & Chan, H.S. 1995 Principles of protein folding – a perspective from simple exact models. *Protein Sci.* (In the press.)

Elöve, G.A., Chaffotte, A.F., Roder, H. & Goldberg, M.E.

1992 Early steps in cytochrome *c* folding probed by time-resolved circular dichroism and fluorescence spectroscopy. *Biochemistry* **31**, 6876–6883.

Eriksson, A.E., Baase, W.A., Zhang, X.J., Heinz, D.W., Blaber, M., Baldwin, E.P. & Matthews, B.W. 1992 Response of a protein structure to cavity-creating mutations and its relation to the hydrophobic effect. *Science, Wash.* **255**, 178–183.

Feng, Y.Q., Wand, A.J. & Sligar, S.G. 1994 Solution structure of apocytochrome b562. *Nature Struct. Biol.* **1**, 30–35.

Fiebig, K.M. & Dill, K.A. 1993 Protein core assembly processes. *J. chem. Phys.* **98**, 3475–3487.

Frauenfelder, H., Petsko, G.A. & Tsernoglou, D. 1979 Temperature-dependent X-ray diffraction as a probe of protein structural dynamics. *Nature, Lond.* **280**, 558–563.

Fujishige, S., Kubota, K. & Anto, I. 1989 Phase transition of aqueous solutions of poly(*N*-isopropylacrylamide) and poly(*N*-isopropylmethacrylamide). *J. phys. Chem.* **93**, 3311–3313.

Gilmanshin, R.I. & Ptitsyn, O.B. 1987 An early intermediate of refolding α-lactalbumin forms within 20 ms. *FEBS Lett.* **223**, 327–329.

Grosberg, A.Yu & Khokhlov, A.R. 1987 Physics of phase transitions in solutions of macromolecules. *Soviet Sci. Rev.* **A8**, 147–258.

Gurd, F.R. & Rothgeb, T.M. 1979 Motions in proteins. *Adv. Protein Chem.* **33**, 73–165.

Hooke, S.D., Radford, S.E. & Dobson, C.M. 1994 The refolding of human lysozyme: a comparison with the structurally homologous hen lysozyme. *Biochemistry* **33**, 5867–5876.

Hughson, F.M., Barrick, D. & Baldwin, R.L. 1991 Probing the stability of a partly folded apomyoglobin intermediate by site-directed mutagenesis. *Biochemistry* **30**, 4113–4118.

Itzhaki, L.S., Evans, P.A., Dobson, C.M. & Radford, S.E. 1994 Tertiary interactions in the folding pathway of hen lysozyme – kinetic studies using fluorescent probes. *Biochemistry* **33**, 5212–5220.

Jennings, P.A. & Wright, P.E. 1993 Formation of a molten globule intermediate early in the kinetic folding pathway of apomyoglobin. *Science, Wash.* **262**, 892–896.

Karplus, M. & McCammon, J.A. 1981 The internal dynamics of globular proteins. *CRC Crit. Rev. Biochem.* **9**, 293–349.

Karplus, M. & Shakhnovich, E. 1992 Protein folding: theoretical studies of thermodynamics and dynamics. In *Protein folding* (ed. T. Creighton), pp. 127–195. New York: Freeman.

Karplus, M. & Weaver, D.L. 1976 Protein-folding dynamics. *Nature, Lond.* **260**, 404–406.

Karplus, M. & Weaver, D.L. 1994. Protein folding dynamics: the diffusion-collision model and experimental data. *Protein Sci.* **3**, 650–668.

Keefe, L.J., Sondek, J., Shortle, D. & Lattman, E.E. 1993 The alpha aneurism: a structural motif revealed in an insertion mutant of staphylococcal nuclease. *Proc. natn. Acad. Sci. U.S.A.* **90**, 3275–3279.

Lattman, E.E., Fiebig, K.M. & Dill, K.A. 1994 Modeling compact denatured states of proteins. *Biochemistry* **33**, 6158–6166.

Lau, K.F. & Dill, K.A. 1989 A lattice statistical mechanics model of the conformational and sequence spaces of proteins. *Macromolecules* **22**, 3986–3997.

Lau, K.F. & Dill, K.A. 1990 Theory for protein mutability and biogenesis. *Proc. natn. Acad. Sci. U.S.A.* **87**, 638–642.

Lifson, S. & Roig, A. 1961 On the theory of helix-coil transition in polypeptides. *J. chem. Phys.* **34**, 1963–1974.

Lim, W.A. & Sauer, R.T. 1991 The role of internal packing interactions in determining the structure and stability of a protein. *J. molec. Biol.* **219**, 359–376.

Lumry, R., Biltonen, R. & Brandts, J.F. 1966 Validity of the 'two-state' hypothesis for conformational transitions of proteins. *Biopolymers* **4**, 917–944.

Matthews, B.W. 1987 Genetic and structural analysis of the protein stability problem. *Biochemistry* **26**, 6885–6888.

Matthews, B.W. 1993 Structural and genetic analysis of protein stability. *A. Rev. Biochem.* **62**, 139–160.

McRee, D.E., Redford, S.M., Getzoff, E.D., Lepock, J.R., Hallewell, R.A. & Tainer, J.A. 1990 Changes in crystallographic structure and thermostability of a Cu, Zn superoxide dismutase mutant resulting from the removal of a buried cysteine. *J. biol. Chem.* **265**, 14234–14241.

Meewes, M., Ricka, J., de Silva, R., Nyffenegger, R. & Binkert, Th. 1991 Coil-globule transition of poly(*N*-isopropylacrylamide). A study of surfactant effects by light scattering. *Macromolecules* **24**, 5811–5816.

Moore, M.A. 1977 Theory of the polymer coil-globule transition. J. Phys. A **10**, 305–314.

Nishii, I., Kataoka, M., Tokunaga, F. & Goto, Y. 1994 Cold denaturation of the molten globule states of apomyoglobin and a profile for protein folding. *Biochemistry* **33**, 4903–4909.

Novotny, J., Bruccoleri, R., Karplus, M. 1984 An analysis of incorrectly folded protein models. Implications for structure predictions. *J. molec. Biol.* **177**, 787–818.

Peng, Z. & Kim, P.S. 1994 A protein dissection study of a molten globule. *Biochemistry* **33**, 2136–2141.

Poland, D.C. & Scheraga, H.A. 1970 *Theory of the helix-coil transition.* New York: Academic Press.

Post, C.B. & Zimm, B.H. 1979 Internal condensation of a single DNA molecule. *Biopolymers* 18, 1487–1501.

Privalov, P.L. 1979 Stability of proteins: small globular proteins. *Adv. Protein Chem.* **33**, 167–241.

Privalov, P.L. & Gill, S.J. 1988 Stability of protein structure and hydrophobic interaction. *Adv. Protein Chem.* **39**, 191–234.

Ptitsyn, O.B., Kron, A.K. & Eizner, Yu. Ye. 1968 The models of the denaturation of globular proteins. I. Theory of globule-coil transitions in macromolecules. *J. Polymer Sci.* C **16**, 3509–3517.

Ptitsyn, O.B., Lim, V.I. & Finkelstein, A.V. 1972 *Analysis and simulation of biochemical systems. Proceedings of the Eighth FEBS Meeting* (ed. B. Hess & H.C. Hemker), pp. 421–431. Amsterdam: North-Holland.

Ptitsyn, O.B. 1987 Protein folding: hypotheses and experiments. *J. Protein Chem.* **6**, 273–293.

Richards, F.M. 1974 The interpretation of protein structures: total volume, group volume distributions and packing density. *J. molec. Biol.* **82**, 1–14.

Richards, F.M. 1977 Areas, volumes, packing and protein structure. *A. Rev. Biophys. Bioeng.* **6**, 151–176.

Richards, F.M. & Lim, W.A. 1994 An analysis of packing in the protein folding problem. *Q. Rev. Biophys.* **26**, 423–498.

Ricka, J., Meewes, M., Nyffenegger, R. & Binkert, Th. 1990 Intermolecular and intramolecular solubilization: collapse and expansion of a polymer chain in surfactant solutions. *Phys. Rev. Lett.* **65**, 657–660.

Sanchez, I.C. 1979 Phase transition behaviour of the isolated polymer chain. *Macromolecules* **12**, 980–988.

Schellman, J.A. 1958 The factors affecting the stability of hydrogen-bonded polypeptide structures in solution. *J. phys. Chem.* **62**, 1485–1494.

Scholtz, J.M., Qian, H., York, E.J., Stewart, J.M. & Baldwin, R.L. 1991 Parameters of helix-coil transition theory for alanine-based peptides of varying chain lengths in water. *Biopolymers* **31**, 1463–1470.

Scholtz, J.M. & Baldwin, R.L. 1992. The mechanism of α-helix formation by peptides. *A. Rev. Biophys. biomolec. Struct.* **21**, 95–118.

Semisotnov, G.V., Rodionova, N.A., Kutyshenko, V.P., Ebert, B., Blanck, J. & Ptitsyn, O.B. 1987 Sequential mechanism of refolding of carbonic anhydrase B. *FEBS Lett.* **224**, 9–13.

Serrano, L., Matouschek, A. & Fersht, A.R. 1992 The folding of an enzyme. III. Structure of the transition state for unfolding of barnase analysed by a protein engineering procedure. *J. molec. Biol.* **224**, 805–818.

Shakhnovich, E.I. & Finkelstein, A.V. 1989 Theory of cooperative transitions in protein molecules. I. Why denaturation of globular protein is a first-order phase transition. *Biopolymers* **28**, 1667–1680.

Shortle, D., Chan, H.S. & Dill, K.A. 1992 Modeling the effects of mutations on the denatured states of proteins. *Protein Sci.* **1**, 201–215.

Shortle, D. 1993 Denatured states of proteins and their roles in folding and stability. *Curr. Opin. Struct. Biol.* **3**, 66–74.

Singh, J. & Thornton, J.M. 1990 SIRIUS. An automated method for the analysis of the preferred packing arrangements between protein groups. *J. molec. Biol.* **211**, 595–615.

Singh, J. & Thornton, J.M. 1992 *Atlas of protein side-chain interactions.* New York: Oxford University Press.

Sosnick, T.R., Mayne, L., Hiller, R. & Englander, S.W. 1994 The barriers in protein folding. *Nature Struct. Biol.* **1**, 149–156.

Stanley, H.E. 1971, 1987 *Introduction to phase transitions and critical phenomena.* New York: Oxford University Press.

Swindells, M.B. & Thornton, J.M. 1993 A study of structural determinants in the interleukin-1 fold. *Protein Engng* **6**, 711–715.

Tiktopulo, E.I., Bychkova, V.E., Ricka, J. & Ptitsyn, O.B. 1994 Cooperativity of the coil-globule transition in a homopolymer – microcalorimetric study of poly(*N*-isopropylacrylamide). *Macromolecules* **27**, 2879–2882.

Uversky, V.N. 1993 Use of fast protein size-exclusion liquid chromatography to study the unfolding of proteins which denature through the molten globule. *Biochemistry* **32**, 13288–13298.

Varadarajan, R. & Richards, F.M. 1992 Crystallographic structures of ribonuclease S variants with nonpolar substitution at position 13: packing and cavities. *Biochemistry* **31**, 12315–12327.

Wuthrich, K. & Wagner, G. 1978 Internal motions in globular proteins. *Trends Biochem. Sci.* **3**, 227–230.

Zimm, B.H. & Bragg, J.K. 1959 Theory of the phase transition between helix and random coil in polypeptide chains. *J. chem. Phys.* **31**, 526–535.

Protein folds: towards understanding folding from inspection of native structures

JANET M. THORNTON, DAVID T. JONES, MALCOLM W. MacARTHUR, CHRISTINE M. ORENGO AND MARK B. SWINDELLS*

Biomolecular Structure and Modelling Unit, Biochemistry and Molecular Biology Department, University College, Gower Street, London WC1E 6BT, U.K.

SUMMARY

Following a short summary of some of the principal features of folded proteins, the results of two complementary studies of protein structure are presented, the first concerned with the factors which influence secondary structure propensity and the second an analysis of protein topology. In an attempt to deconvolute the physical contributions to secondary structure propensities, we have calculated intrinsic ϕ,ψ propensities, derived from the coil regions of proteins. Comparison of intrinsic ϕ,ψ propensities with their equivalent secondary structure values show correlations for both helix and strand. This suggests that the local dipeptide, steric and electrostatic interactions have a major influence on secondary structure propensity. We then proceed to inspect the distribution of protein domain folds observed to date. Several folds occur very commonly, so that 46% of the current non-homologous database comprises only nine folds. The implications of these results for protein folding are discussed.

1. INTRODUCTION: A SUMMARY OF SOME MAJOR DETERMINANTS OF PROTEIN FOLDS AND FOLDING

It is not straightforward to infer information about the folding pathway from an inspection of the final native state of the protein. It is somewhat analogous to the problem facing astrophysicists, who must deduce information about the origins of the universe from the current state of the galaxies. Fortunately we are able to replay folding many times experimentally, but it is still important to use all the information that has been collected on protein structures to improve our understanding of the folding process. Before describing the results of two detailed studies we have recently completed at University College London, it is appropriate to review some of the basic principles which emerge from an inspection of the many structures in the Protein Structure Databank (PDB; Bernstein *et al.* 1977). In this first section we shall highlight principles drawn from work in our laboratory over the past few years, as well as many other groups.

Five essential observations can be made from inspection of protein structures.

1. All proteins exhibit a tightly packed hydrophobic core (Richards 1977; Hubbard *et al.* 1994). In a recent inspection of high resolution structures, Williams *et al.* (1994) found that on average water-sized cavities

constitute only 1% of the volume of a protein. The small number of buried waters (about one per 27 residues on average) and cavities suggests that close packing, exclusion of water and burial of hydrophobic groups are major determinants of protein folding.

2. In X-ray derived structures, determined to high resolution, the ϕ,ψ and χ torsion angles are generally confined to low energy conformations (Morris *et al.* 1991). Indeed in such structures almost 90% of ϕ,ψ angles lie in only 14% of ϕ,ψ space. There are some well documented examples of distorted torsion angles (Herzberg & Moult 1991), but these are the exception rather than the rule. Thus, even in the complex interior of a folded structure, the local interactons are sufficiently strong to provide powerful restraints on torsional freedom. During folding, in the absence of strong tertiary interactions, these torsional angles are even more likely to adopt their preferred low energy conformers.

3. Potential hydrogen bond donors and acceptors are nearly always satisfied (McDonald & Thornton 1994). Only about 2% of main chain carbonyls and 6% of main chain amide groups fail to form hydrogen bonds to the protein or the solvent. Thus satisfaction of hydrogen bond potential is clearly an important constraint, which will also apply during folding. It is obviously energetically expensive to bury a potential donor or acceptor without satisfying its potential. As with the torsion angles, there are some exceptions, but these are rare and presumably must be compensated by other favourable interactions. It is this effect which makes the formation of secondary structure an obliga-

---

* Present address: Department of Molecular Design, Yamanouchi Pharmaceutical Co. Ltd., 21 Miyukigaoka, Tsukaba 305, Tokyo, Japan.

Phil. Trans. R. Soc. Lond. B (1995) **348**, 71–79
Printed in Great Britain

71

© 1995 The Royal Society

tory feature of compact globular structures, where the main chain is buried in the interior. Only by formation of the regular hydrogen bonds seen in sheets and helices can all the main chain groups be satisfied.

4. Side chain packing varies from random to rather specific, depending on the type of amino acids involved (Singh & Thornton 1990). Interactions between oppositely charged residues (e.g. arginine and aspartic acid) lead to definite preferred orientational patterns, which are clearly seen in proteins. The polar interactions also confer orientational preferences, although these are not so strong as those seen for the formally charged groups. In contrast the apolar interactions appear to be essentially randomly distributed in space. Provided these side chains are shielded from solvent and reasonably well packed they do not have any preferred spatial orientations. Interestingly, the aromatic ring sidechains behave more like polar residues than apolar. There are distinct preferences for negatively charged polar atoms (e.g. oxygens and sulphurs) to eschew the electronegative face of the aromatic ring and prefer to pack against the positively charged edge of the ring (Reid *et al.* 1985). Stacking interactions between aromatics are relatively rare, whereas the energetically more favourable edge–face interactions are much more common (Burley & Petsko 1985; Singh & Thornton 1985). Thus the interactions between side chains are not random, except for the apolar–apolar contacts, which may well dominate the core of the protein. However, specific packing will result from specific polar interactions especially hydrogen bonds, and these may well be vital in the folding process.

5. Protein structures are dominated by the secondary structures adopted on average by 60% of all residues. Hydrogen bonding interactions between strands dictate strand geometry. Similarly specific helix–helix and sheet–sheet interactions are favoured as determined by the requirement for close packing (Chothia & Finkelstein 1990).

These observations reveal much about the energetic forces which drive folding. Folding can be seen as a balance between satisfying the local conformational preferences and the global requirement to bury apolar sidechains, whilst satisfying almost all potential hydrogen bond donors and acceptors. Below we explore two aspects of protein structure with their implications for folding in more detail. First we analyse the preferred ϕ,ψ conformations for the 20 amino acids and then we consider the distribution of currently known structures among the possible protein topologies.

2. SECONDARY STRUCTURE PROPENSITIES

To facilitate a more thorough understanding of secondary structure propensities, we have developed a novel procedure for deconvoluting the competing factors which govern secondary structure formation. Our work is statistically based, but differs from previous reports (Chou & Fasman 1978) because we calculate the ϕ,ψ preferences in coil regions and separate them from all other structural data (figure

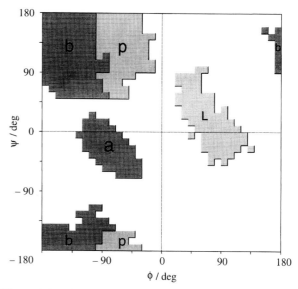

Figure 1. Digitized plot showing the ϕ,ψ regions used in this work: a, b and p. These follow the definitions of Efimov (1980) and Wilmot & Thornton (1990). The b and p regions collectively constitute the B region. Delineation of b and p regions is important because the fixed ϕ angle essentially excludes Pro from the b region. In contrast, other residues can occupy both the b and p regions, although strand residues prefer the b region. For completeness the aL and gL regions are also shown collectively (L).

1*a, b*). By omitting regular interactions from residues located in helices and strands, it is possible to calculate intrinsic preferences for specific regions of ϕ,ψ space. We refer to these regions as a, b, p and B (see figure 1).

(a) *Methods and data*

Using a dataset of 85 structures from the PDB, propensities for a/coil, b/coil, p/coil, B/coil, helix and strand were calculated (table 1).

Using alanine in the a/coil state as an example, intrinsic propensities for the coil state were calculated in the following manner.

$n(\text{Ala})_{\text{a/coil}}$ = number of alanine residues adopting an a region conformation when in coil,

$n(\text{Ala})_{\text{coil}}$ = total number of alanine residues in coil,

$N_{\text{a/coil}}$ = total number of residues in coil,

N_{coil} = number of residues adopting an a region conformation when in coil,

$P(\text{Ala})_{\text{a/coil}} = \{n(\text{Ala})_{\text{a/coil}}/n(\text{Ala})_{\text{coil}}\}/\{N_{\text{a/coil}}/N_{\text{coil}}\}$,

where $P(\text{Ala})_{\text{a/coil}}$ is the propensity for Ala to adopt the a/coil conformation.

In this manner, $P(\text{Ala})_{\text{a/coil}}$ measures the propensity for alanine to adopt a ϕ,ψ conformation within the a region, given that it is in the coil state. It gives no indication of the relative preferences for coil, strand and helix. The regular secondary structure propensities are calculated following the standard formalism of Chou & Fasman (1978) using the secondary structure definitions of Kabsch & Sander (1983).

For the analysis of χ_1 angles, we use the formalism: gauche plus $(\text{g}^+) = -60°$, trans $(\text{t}) = 180°$ and gauche minus $(\text{g}^-) = +60°$. A correction is applied for

Table 1. *Intrinsic propensities for a/coil, b/coil, p/coil, B/coil, and Chou & Fasman type propensities for α-helix and β-strand*

| acid | intrinsic φ,ψ propensities | | | | | regular secondary structure propensities | |
	a/coil	b/coil	p/coil	B/coil	other	α-helix	β-strand
Gly	0.33	0.34	0.31	0.32	3.80	0.41	0.64
Ala	1.23	0.78	1.32	1.09	0.39	1.47	0.79
Val	0.89	1.83	0.96	1.33	0.36	0.95	1.73
Leu	1.16	0.82	1.40	1.15	0.35	1.32	1.17
Ile	0.98	1.68	0.99	1.29	0.32	1.13	1.76
Phe	0.93	1.63	0.93	1.23	0.55	1.04	1.39
Tyr	0.85	1.46	1.12	1.26	0.59	0.88	1.52
Trp	1.17	0.90	1.24	1.09	0.48	1.05	1.25
Pro	1.00	0.10	2.29	1.35	0.14	0.46	0.42
Cys	0.87	1.34	1.32	1.33	0.41	0.89	1.18
Met	1.07	1.05	1.23	1.15	0.51	1.37	1.32
Ser	1.29	0.95	1.00	0.98	0.56	0.71	0.93
Thr	1.13	1.39	0.96	1.15	0.43	0.71	1.27
Lys	1.20	1.07	0.94	0.99	0.68	1.10	0.92
Arg	1.09	1.40	0.74	1.03	0.77	1.41	0.71
His	0.93	1.37	0.84	1.07	0.95	0.97	0.86
Asp	1.16	1.18	0.82	0.98	0.80	0.85	0.49
Asn	0.79	1.35	0.60	0.92	1.54	0.78	0.56
Glu	1.45	0.84	0.95	0.90	0.50	1.39	0.78
Gln	1.26	1.07	1.00	1.03	0.48	1.36	0.81

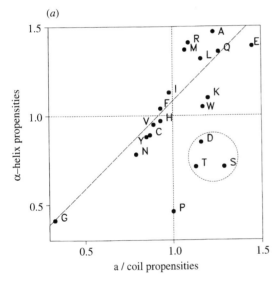

(a)

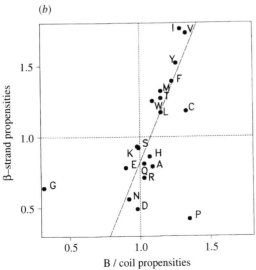

(b)

Figure 2. Graphs showing the propensities for (a) a/coil versus helix and (b) B/coil versus strand.

the anomaly in valine χ_1 classification. Thus valine (t, g⁻, g⁺) wells are listed as (g⁺, t, g⁻) in this paper. Calculations of χ_1 angle propensities within each region of the φ,ψ plot are implemented in a similar manner to those above. For example, the propensity of threonine to adopt the g⁺ conformation, given that it is both in the a region and coil state can be described as:

$$P(\text{Thr})_{g+/\text{acoil}}$$
$$= \{n(\text{Thr})_{g+/\text{acoil}}/n(\text{Thr})_{\text{acoil}}\}/\{N_{g+/\text{acoil}}/N_{\text{acoil}}\}.$$

Glycine and alanine, which do not have side chains, have been omitted.

(b) Results

It is clear that each residue type has intrinsic propensities for different regions of φ,ψ space, and that these values do not necessarily agree with the associated secondary structure propensities. In figure 2a, b, comparisons are made between the propensities for a/coil and helix, as well as B/coil and strand. Although it is not meaningful to compare the absolute values of the residue propensities with one another (because the a/coil and B/coil values are only calculated from the coil subset), relative comparisons can be made. The correlation between a/coil and helix values is relatively weak; Pearson correlation coefficients are 0.60 for all residues and 0.44 when Gly and Pro are excluded. In addition, several residues, especially Ser, Thr and Asp show markedly different rank ordering. If Ser, Thr, Asp, Gly and Pro are all excluded, the correlation coefficient increases to 0.81. In contrast, although the correlation coefficient between B/coil and strand propensities is low when all residues are considered (correlation coefficient = 0.53) it becomes much higher when Gly and Pro are excluded (correlation coefficient = 0.86).

These data suggest that the intrinsic ϕ,ψ preferences, and their compatibility with the observed secondary structure propensities, vary with the side-chain concerned. Intrinsic preferences must principally reflect side-chain interactions with the two local peptide units (Ralston & DeCoen 1974; Finkelstein & Ptitsyn 1976a, b; Zimmerman et al. 1977) and these should be evident in the χ_1 distributions. As expected from simple steric considerations, χ_1 values fall into three preferred zones ($+60°$ g$^-$, $180°$ trans, $-60°$ g$^+$) (Janin et al. 1978; McGregor et al. 1987; Ponder & Richards 1987; Dunbrack & Karplus 1994). Comparisons of the χ_1 preferences in a/coil and B/coil regions, are significantly different from those occuring in the equivalent regular secondary structures (see table 2).

How then can we rationalize these variations? Glu, Gln, Ser and Asp all have high a/coil propensities (figure 2a) and have a polar or charged oxygen acceptor, which can interact favourably with the main-chain NH groups. It is not surprising therefore that Asp and Ser χ_1 angles strongly favour the g$^-$ state, as this conformation facilitates the electrostatic interaction (table 2). This preference is not observed for Glu and Gln because the long, flexible side-chain can form electrostatic interactions in other χ_1 conformers as well. Thr has a slightly lower a/coil propensity than Ser, even though it can stabilize the a/coil conformation via its hydroxyl, and has a χ_1 distribution which strongly favours the required g$^+$ conformer. This can be attributed to the branched C$^\beta$ atom which causes steric clashes, similar to those observed in Val and Ile (see below). Consequently, the a/coil propensity is intermediate between the polar Ser and apolar Val and Ile.

Although Ser, Asp and Thr have high a/coil propensities, their helix propensities are low. This is because the g$^-$ rotamer which stabilizes the a/coil conformation, is effectively forbidden in a helix, as the side-chain will interfere with the hydrogen bonds required for helix formation. In contrast, the side-chains of Glu and Gln are presumably sufficiently long and flexible to be displaced as the helix is formed. As a result, their high a/coil propensities are translated into high propensities for helix formation.

Both Leu and Ala exhibit reasonably high a/coil values and very high helix propensities. The most noticeable property of Leu is that, even in the a/coil conformation its χ_1 angle rarely adopts the g$^-$ conformation, due to side-chain main-chain steric clashes (table 1). Because the g$^-$ state is usually forbidden in helices anyway, due to unfavourable interactions with the previous helical turn, Leu can convert from a/coil to helix without the normal loss of side-chain entropy. Leu and Ala (which has no C$^\gamma$) are the only two amino acids for which this observation holds, and both have very high helix propensities. One other striking observation is that Leu and Ala are the only two amino acids which have a really strong preference for the p/coil region of ϕ,ψ space, rather than the b/coil region. This may also influence the strong helix propensities observed for Ala and Leu, as only a ψ angle rotation is required to move from the p region to the helix-forming a region.

Ile and Val, both have very low a/coil and helix propensities, and clearly cannot provide any electrostatic stabilization for the a/coil conformation. Inevitably this leads to a preference for B/coil. Within the B/coil region, Ile and Val both have a strong preference for b/coil, due to steric effects (between NH and C$^\gamma$ atoms) which occur in the p/coil region when residues have a branched C$^\beta$ atom. As a result, the preference for strand formation rather than helix is further enhanced. Although non-polar residues cannot stabilize the a/coil conformation, and in general have a low propensity for the 'aligned' dipole conformation in a/coil, their presence in helices is often obligatory, since helices require a hydrophobic face which can pack against the rest of the hydrophobic core.

As the correlation between B/coil and strand propensities is high (0.86 when Pro and Gly are omitted), it would appear that the intrinsic propensities for B/coil are a major factor in determining those for strand formation. In general, hydrophobic amino acids have the highest B/coil propensities. Of the polar residues, positively charged side-chains such as Arg, Lys and His have slightly higher B/coil propensities than Ser, Asp, Asn and Glu, although the differences are not pronounced. These elevated propensities for positively charged side-chains may reflect their ability to interact favourably with the main-chain CO of residue $i+1$, whereas the lower values for negatively charged residues will inevitably result from their high a/coil propensities. In an extended conformation there are no interactions between sequential side-chains, and their peptide groups lie in a plane which is almost perpendicular to the side-chain C$^\alpha$–C$^\beta$ bond. Thus there is also little interaction between side-chain and main-chain. This allows hydrophobic side-chains to be buried and main-chain polar groups to have their hydrogen bonding potentials satisfied through solvation.

(c) Implications for folding

What are the implications for folding of these empirical ϕ,ψ propensities, which derive from local interactions between the side chain and main chain atoms. In the initial stages of folding, the polypeptide chain will adopt a random coil conformation. However, even at this stage the sequence will influence the coil conformation adopted as driven by the intrinsic ϕ,ψ propensities. Thus Leu will predominantly adopt the a/coil conformer, whereas Ile and Val will prefer the b/coil state. As the chain folds the medium range interactions ($i \ldots i+3,4$) which occur in a helix will come into effect, modulating the intrinsic ϕ,ψ propensities. Thus Leu, which usually adopts the g$+\chi_1$ conformer favoured in a helix, is most readily 'accepted' into a helix, whereas Ser and Thr have difficulty in forming a helix because of the local side chain–backbone interactions. Similarly, the branched apolar amino acids are further discriminated against by the helix geometry. The close contacts in a helix normally distort the χ_1 angle to alleviate the strain, but this is not possible for the branched C$^\beta$ sidechains.

As well as influencing secondary structure formation,

Table 2. χ_1 angle distributions and propensities for residues in different regions of ϕ, ψ space and secondary structure

(Distributions are shown in normal type and propensities in bold. Ala and Gly have no χ angles and those of Pro are atypical. For χ_1 angles we use the formalism: gauche plus(g$^+$) = 60°, trans(t) = 180° and gauche minus(g$^-$) = +60°. A correction is applied for the anomaly in valine χ_1 classification. Thus, valine(t, g$^-$, g$^+$) wells are listed as (g$^+$, t, g$^-$) in this paper. Calculations of χ_1 angle propensities within each region of the ϕ, ψ plot are implemented in a similar manner to those in table 1. For example, the propensity for threonine to adopt the g$^+$ conformation, given that it is both in the **a** region and coil, is described as: $P(\mathrm{Thr})_{\mathrm{g+/acoil}} = \{n(\mathrm{Thr})_{\mathrm{g+/acoil}}/n(\mathrm{Thr})_{\mathrm{acoil}}\}/(N_{\mathrm{g+/acoil}}/N_{\mathrm{acoil}}).$)

amino acid	a/coil			b/coil			p/coil			B/coil (b+p)			α-helix			β-strands		
	g⁻	t	g⁺	g⁻	t	g⁺	g⁻	t	g⁺	g⁻	t	g⁺	g⁻	t	g⁺	g⁻	t	g⁺
Val	46 **1.24**	20 **1.42**	48 **0.77**	58 **1.45**	17 **0.40**	73 **1.11**	26 **1.60**	8 **0.24**	69 **1.28**	84 **1.57**	25 **0.33**	142 **1.17**	32 **1.27**	16 **0.15**	301 **1.38**	100 **1.19**	42 **0.29**	350 **1.32**
Leu	1 **0.01**	30 **1.39**	143 **1.49**	2 **0.09**	15 **0.67**	61 **1.77**	1 **0.04**	48 **0.85**	127 **1.38**	3 **0.06**	63 **0.81**	188 **1.53**	1 **0.03**	190 **1.18**	344 **1.03**	13 **0.21**	171 **1.61**	181 **0.92**
Ile	35 **1.47**	16 **1.77**	22 **0.55**	25 **1.17**	5 **0.22**	49 **1.40**	8 **0.82**	7 **0.35**	47 **1.45**	33 **1.10**	12 **0.28**	96 **1.41**	19 **0.88**	18 **0.20**	261 **1.40**	46 **0.75**	43 **0.41**	269 **1.39**
Phe	7 **0.28**	12 **1.26**	58 **1.37**	16 **0.69**	19 **0.77**	51 **1.34**	3 **0.29**	19 **0.91**	43 **1.27**	19 **0.59**	38 **0.83**	94 **1.29**	3 **0.19**	131 **2.04**	79 **0.59**	54 **1.44**	52 **0.82**	113 **0.96**
Tyr	10 **0.42**	13 **1.44**	50 **1.25**	14 **0.66**	24 **1.06**	41 **1.17**	7 **0.55**	42 **1.64**	31 **0.74**	21 **0.62**	66 **1.36**	72 **0.94**	6 **0.47**	98 **1.85**	72 **0.65**	48 **1.20**	62 **0.91**	125 **0.99**
Trp	8 **0.66**	4 **0.87**	25 **1.23**	5 **1.03**	6 **1.16**	7 **0.88**	2 **0.38**	15 **1.42**	16 **0.93**	7 **0.64**	21 **1.35**	23 **0.93**	7 **1.27**	39 **1.71**	29 **0.61**	16 **1.34**	15 **0.74**	39 **1.03**
Pro	117 **1.83**	0 **0.00**	79 **0.73**	11 **3.40**	0 **0.00**	0 **0**	187 **3.16**	0 **0.00**	188 **0.96**	198 **2.40**	0 **1.00**	188 **1.01**	30 **3.92**	0 **0.00**	76 **1.14**	39 **3.04**	0 **0.00**	36 **0.89**
Cys	12 **0.85**	1 **0.19**	30 **1.27**	14 **1.23**	14 **1.16**	14 **0.75**	2 **0.23**	21 **1.19**	32 **1.12**	16 **0.77**	35 **1.19**	46 **0.98**	5 **0.80**	28 **1.07**	54 **0.99**	10 **0.66**	34 **1.32**	45 **0.94**
Met	4 **0.38**	3 **0.76**	25 **1.42**	5 **0.93**	6 **1.04**	9 **1.02**	1 **0.20**	11 **1.11**	19 **1.18**	6 **0.55**	17 **1.10**	28 **1.14**	3 **0.32**	34 **0.87**	92 **1.13**	15 **0.90**	29 **1.03**	53 **1.01**
Ser	187 **2.05**	12 **0.35**	81 **0.53**	61 **1.74**	45 **1.20**	23 **0.40**	63 **2.19**	71 **1.22**	47 **0.50**	124 **1.87**	116 **1.22**	70 **0.46**	86 **4.86**	53 **0.72**	106 **0.69**	88 **2.08**	89 **1.24**	71 **0.53**
Thr	152 **2.54**	3 **0.13**	27 **0.27**	86 **2.23**	20 **0.49**	37 **0.58**	57 **2.76**	9 **0.21**	65 **0.95**	143 **2.45**	29 **0.35**	102 **0.77**	50 **3.28**	6 **0.09**	155 **1.17**	78 **1.56**	39 **0.46**	175 **1.11**
Lys	19 **0.32**	34 **1.53**	119 **1.20**	14 **0.51**	21 **0.72**	64 **1.42**	8 **0.43**	37 **0.98**	71 **1.15**	22 **0.47**	58 **0.87**	135 **1.27**	21 **0.88**	135 **1.36**	168 **0.81**	19 **0.53**	94 **1.53**	98 **0.86**
Arg	11 **0.31**	26 **1.96**	70 **1.19**	12 **0.51**	23 **0.92**	52 **1.35**	5 **0.52**	19 **0.97**	37 **1.16**	17 **0.54**	42 **0.93**	89 **1.25**	10 **0.46**	149 **1.64**	142 **0.75**	20 **0.99**	35 **1.02**	62 **0.98**
His	16 **0.83**	8 **1.10**	35 **1.08**	9 **0.61**	13 **0.82**	33 **1.36**	6 **0.85**	19 **1.32**	20 **0.85**	15 **0.70**	32 **1.05**	53 **1.10**	8 **1.01**	46 **1.39**	56 **0.81**	15 **1.17**	26 **1.19**	34 **0.84**
Asp	79 **1.03**	11 **0.38**	142 **1.10**	26 **0.63**	103 **2.36**	23 **0.34**	22 **1.00**	67 **1.49**	51 **0.70**	48 **0.77**	170 **1.91**	74 **0.53**	11 **0.61**	40 **0.53**	197 **1.26**	11 **0.57**	65 **2.00**	36 **0.60**
Asn	39 **0.90**	9 **0.55**	84 **1.15**	33 **0.85**	73 **1.76**	37 **0.58**	16 **1.21**	35 **1.30**	32 **0.73**	49 **1.01**	108 **1.56**	69 **0.63**	4 **0.29**	32 **0.57**	152 **1.29**	9 **0.51**	51 **1.71**	43 **0.77**
Glu	39 **0.64**	33 **1.43**	112 **1.10**	12 **0.65**	12 **0.61**	43 **1.43**	11 **0.68**	34 **1.04**	54 **1.01**	23 **0.64**	46 **0.89**	97 **1.18**	21 **0.73**	132 **1.10**	240 **0.96**	19 **0.64**	80 **1.58**	72 **0.77**
Gln	11 **0.33**	21 **1.68**	67 **1.21**	7 **0.48**	9 **0.58**	37 **1.55**	3 **0.28**	28 **1.30**	36 **1.03**	10 **0.39**	37 **1.00**	73 **1.25**	11 **0.63**	93 **1.28**	137 **0.90**	14 **0.74**	50 **1.55**	47 **0.79**

the intrinsic ϕ,ψ propensities also help to determine the conformation of the coil regions. Folding can be seen as a balance between satisfying these local conformational preferences and the global requirement to bury apolar sidechains, while satisfying almost all potential hydrogen bond donors and acceptors.

3. PROTEIN FOLDS: OBSERVED DISTRIBUTION OF PROTEINS AMONG THE FOLD FAMILIES

Having assessed the influence of local interactions on secondary structure and folding, we now consider tertiary structure and an analysis of the global topologies observed for polypeptide chains. It is well established that protein domains, having more than 30% of their amino acid sequences in common, will adopt the same three-dimensional structure. The core of these structures is usually well conserved, whereas their surfaces may incorporate many differences in loop insertions or deletions. Furthermore, as the number of known structures increases, it is apparent that some structures adopt the same fold, even though their sequences are apparently completely dissimilar. We wished to consider these observations in more detail, especially with respect to the implications for the folding process. Therefore we have clustered all the protein structures in the PDB, using an automated procedure to compare all structures, which calculates a quantitative measure of their structural similarity.

(a) Comparison of protein structures

All the proteins in the April 1993 PDB including prereleases were clustered using the method of Orengo *et al.* (1993), which includes sequence and structure comparisons. Initially all the protein sequences are compared using the standard Needleman and Wunsch algorithm (Needleman & Wunsch 1970) and similarity is measured by % sequence identity. The protein sequences are then clustered into families which show a clear sequence relationship (more than 30% sequence identity to at least one other member of the family). Representative structures of each family are then compared using the SSAP algorthm (Taylor & Orengo 1989), which finds similar regions by comparing the structural environment of each residue. The optimal alignment is achieved by a complex double dynamic programming process, which allows for insertions and deletions between the structures. The program returns a SSAP score, which is normalized to be in the range 0–100 independent of the protein size. The proteins are then clustered into structural families, according to their SSAP scores. Two structures were deemed to adopt the same fold if their SSAP score was > 70 and, to ensure global similarity, at least 70% of the larger protein must be equivalenced against the smaller. It is important to realize that, as with sequence similarities, structural similarities form a continuum and the cutoffs used are necessarily arbitrary. These cutoffs were chosen to reflect the consensus view of when two structures are similar, as found in the literature.

(b) Homologous and analogous folds

In comparing structures it became clear that there were two situations in which we found that a pair of structures looked similar. In the first, the two proteins have the same fold, and similar functions, even though their sequences are dissimilar. For example haemoglobin (PDB code-1eca) and myoglobin (1mbd) have a SSAP score = 85 when compared, even though their sequences show only 15% identity. It is generally agreed that these proteins have arisen by divergent evolution from a common ancestor, and can be considered to be homologous and belong to the same superfamily. When the structures of such proteins were compared we found that their comparison always yielded a SSAP score > 80. In clustering the databank we wished to signify that these structures are homologous, and therefore consider that they belong to the same hyperfamily (i.e. the same superfamily, but extended to include functionally and structurally similar proteins). In contrast there are other examples where two proteins have a similar structure, yet show no sequence or functional similarity. For example, haemoglobin (1eca) and colicin (1colA) have a sequence identity of 5%, no functional similarity (colicin is a protein which penetrates cell membranes and ultimately causes cell death) and yet have the same protein fold and give a SSAP score = 75. Such similarities will be described as analogous folds, and may have arisen by divergent or convergent evolution. Empirically we have found that these proteins yield a SSAP score > 70 but < 80, being in general less similar than functionally related proteins.

(c) Protein superfamilies and fold families

Thus the proteins in the PDB were clustered by their sequences and structures. Sequence analysis allowed recognition of the superfamilies. These families were then expanded by structural and functional comparisons to include proteins which have the same fold and similar functions, which would generally be considered to be homologous. These clusters were called hyperfamilies. The hyperfamilies could then be clustered into fold families, bringing together proteins which have the same fold but a different function and no obvious sequence similiarity. Thus we have the following hierarchy in the clustering procedure: (i) superfamilies, recognized only from sequence (> 30% identity); (ii) hyperfamilies, very similar structures with SSAP score > 80 and functional similarity; and (iii) fold families, similar folds with SSAP score > 70 but different function.

Using these criteria all the single domain proteins in the PDB were clustered into 392 superfamilies, 274 hyperfamilies and 206 fold familes, as illustrated in figure 3. From this analysis it is apparent that although the PDB is growing rapidly, we still only have a limited number of 'independent' protein structures and still fewer unique domain folds. Furthermore the distribution of the structures between the different fold families is shown in figure 4. This plot is far from a random distribution (Orengo *et al.* 1994). We observe

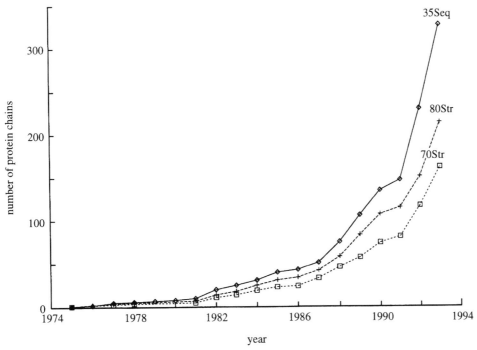

Figure 3. Increase in the number of protein structures deposited each year in the PDB, grouped by sequence family (35SEQ); hyperfamily family (SSAP score > 80) and fold family (SSAP score > 70).

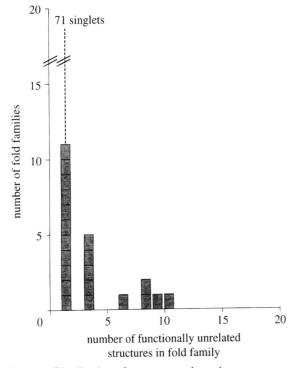

Figure 4. Distribution of current non-homologous structures (< 25 % sequence identity and < 70 SSAP score and no functional similarity) among the different folds. The nine superfolds contain between 3–10 representatives, whereas all the remaining folds are singlets.

nine fold families, for which there are between 3–11 examples and 71 folds for which there is only one singlet example. These very common folds, which are termed superfolds, represent 46 % of all non-homologous proteins in the PDB.

The observation that all folds are not equally populated is not new (Ptitsyn & Finkelstein 1980;

Finkelstein & Ptitsyn 1987). For example, in the four-helix bundle proteins, the up-down-up-down topology occurs frequently, whereas the more complex up-up-down-down fold is relatively rare. Similarly in the beta-sandwich structures, the immunoglobulin fold is very common. In contrast many other folds have only been observed once, although they were among the first folds to be determined (i.e. all examples of the singlet folds in the database are clearly related, as shown either by sequence or similarity of structure and function, e.g. lysozyme and ribonuclease A).

The existence of superfolds has several possible implications for proteins folding. These folds may represent extra stable folds, which have diverged from a common ancestor and, despite extensive changes to their sequences, are resistant to changes in topology. Any residual sequence patterns are commonly undetectable and the 'original' function may have changed. Alternatively these superfolds may reflect the accidental convergence of disparate protein chains to the same topology and will have diverse functions. Regardless of the origins of these folds, it is apparent that they are compatible with a much larger set of sequences, compared to one of the singlet folds (Finkelstein 1994).

The superfolds are shown in figure 5 and it is immediately apparent that they all exhibit simple topologies, with a high percentage of sequential secondary structures which lie adjacent in the tertiary fold. Thus it is tempting to suggest that these proteins are so common because they can fold up more easily or rapidly than other more complex topologies. In an all sequential structure, such as a TIM barrel, there are in principle no requirements for an obligate pathway, as the structure could nucleate anywhere and fold up in any order. If the relative stability of any part of the structure were altered by mutation, thus affecting the

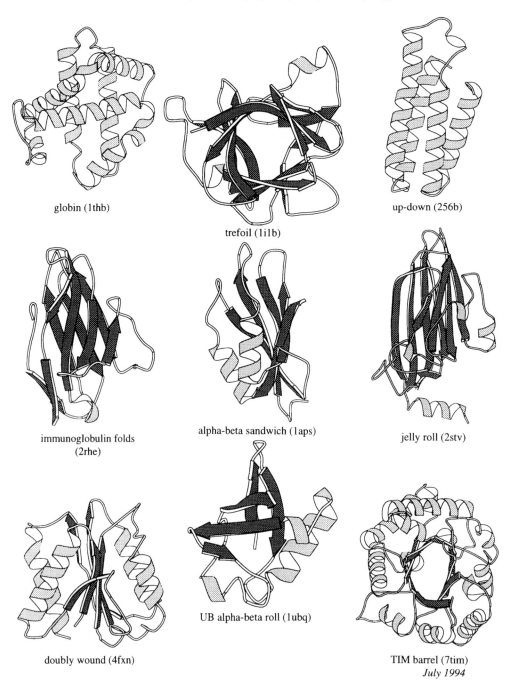

globin (1thb)

trefoil (1i1b)

up-down (256b)

immunoglobulin folds
(2rhe)

alpha-beta sandwich (1aps)

jelly roll (2stv)

doubly wound (4fxn)

UB alpha-beta roll (1ubq)

TIM barrel (7tim)
July 1994

Figure 5. The nine superfold structures. The specific protein shown is indicated in brackets by the PDB code.

local rate of folding, this need not affect the final structure, just the order of folding. If this hypothesis were correct we would expect that the all sequential structures observed to date (the up-down barrels, the β-propeller structures and the β-solenoid fold) would be common. The recently observed 'holy' proteins (Dijkstra & Thunnissen 1994; Hofsteenge 1994) are also all-sequential.

Many of the singlet folds are stabilized by disulphide bridges, and denature if the disulphides are reduced. This adds sensitive hot-spots to the structure, with the requirement for conservation of the disulphide, if the native fold is to be maintained. Without these additional constraints these folds may not be sufficiently stable to withstand random mutagenesis. Members of these fold families can often be identified as related as the pattern of cystines is conserved.

4. CONCLUSION

In this paper we have concentrated on two aspects of protein folding, from the details of the ϕ, ψ distribution to the gross topology of a protein structure. As we discover more about the details of protein structures at all levels of the structural hierarchy, we are better able to understand the folding process and make hypotheses which can be tested experimentally. Clearly the folds we observe reflect the complex interplay of the weak non-covalent interactions between the side chains and the intrinsic conformational preferences of the 20 amino acids in a water environment. It is still not clear whether we need to understand the folding pathway in order to be able to predict protein structure from sequence. However, with current progress in determining folding pathways, it may soon be possible to understand the folding process in much more detail and perhaps model this complex macromolecular process and predict its endpoint.

REFERENCES

Bernstein, F.C., Koetzle, T.F., Williams, G.J.B., Meyer, E.F. Jr, Brice, M.D., Rodgers, J.R., Kennard, O., Shimanouchi, T. & Tasumi, M. 1977 The Protein Data Bank: a computer based archival file for macromolecular structures. *J. molec. Biol.* **122**, 535–542.

Burley, S.K. & Petsko, G.A. 1985 Aromatic-aromatic interactions: a mechanism of protein structure stabilisation. *Science,Wash.* **229**, 23–29.

Chothia, C. & Finkelstein, A. 1990 Tha classification and origins of protein folding patterns. *A. Rev. Biochem.* **59**, 1007–1039.

Chou, P.Y. & Fasman, G.D. 1978 Predictions of the secondary structure of proteins from their amino acid sequence. *Adv. Enzmol.* **47**, 45–148.

Dijkstra, B.W. & Thunnissen, A.-M.W.H. 1994 'Holy' proteins II: the soluble lytic transglcosylase. *Curr. Opin. Struct. Biol.* **4**, 810–813.

Dunbrack, R. & Karplus, M. 1994 Conformational analysis of the backbone dependent rotamer preferences of protein side-chains. *Nature Struct. Biol.* **1**, 334–339.

Efimov, A.V. 1980. Standard conformations of polypeptide chains in irregular regions of proteins. *Molec. Biol. Moscow* **20**, 208–216.

Finkelstein, A.V. & Ptitsyn, O.B. 1976*a* Theory of protein molecule organisation. I. Thermodynamic parameters of local secondary structures in the unfolded protein chain. *Biopolymers* **16**, 469–495.

Finkelstein, A.V. & Ptitsyn, O.B. 1976*b* A theory of protein molecule self organisation. IV. Helical and irregular local structures of unfolded protein chains. *J. molec. Biol.* **103**, 15–24.

Finkelstein, A.V. & Ptitsyn, O.B. 1987 Why do globular proteins fit the limited set of folding patterns. *Prog. Biophys. molec. Biol.* **50**, 171–190.

Herzberg, O. & Moult, J. 1991 Analysis of the steric strain in the polypeptide backbone of protein molecules. *Proteins* **11**, 223–229.

Hofsteenge J. 1994 'Holy' proteins I: ribonuclease inhibitor. *Curr. Opin. Struct. Biol.* **4**, 807–809.

Hubbard, S.J., Gross, K.H. & Argos, P. 1994 Intramolecular cavities in globular proteins. *Protein Eng.* **7**, 613–626.

Janin, J., Wodak, S., Levitt, M. & Maigret, B. 1978 Conformation of amino acid side-chains in proteins. *J. molec. Biol.* **125**, 357–386.

Kabsch, W. & Sander, C. 1983 Dictionary of protein secondary structure. *Biopolymers* **22**, 2577–2637.

McDonald, I.K. & Thornton, J.M. 1994 Satisfying hydrogen bonding potential in proteins *J. molec. Biol.* **238**, 777–793.

McGregor, M.J., Islam, S.A. & Sternberg, M.J.E. 1987 Analysis of the relationship between side-chain conformation and secondary structure in globular proteins. *J. molec. Biol.* **198**, 295–310.

Morris, A.L., MacArthur, M.W., Hutchinson, E.G. & Thornton, J.M. 1992 Stereochemical quality of protein structure coordinates. *Proteins* **12**, 345–364.

Needleman, S.B. & Wunsch, C.D. 1970 A general method applicable to the search for similarities in the amino acid sequences of two proteins. *J. molec. Biol.* **48**, 433.

Orengo, C.A. & Taylor, W.R. 1993 A local alignment method for protein structure motifs. *J. molec. Biol.* **233**, 488–497.

Orengo, C.A., Flores, T.P., Taylor, W.R & Thornton, J.M. 1993 Identification and classification of protein fold families. *Protein Eng.* **6**, 485–500.

Orengo, C.M., Jones, D.T. & Thornton, J.M. 1994 Protein superfamilies and domain superfolds. *Nature, Lond.* **372**, 631–634.

Ponder, J.W. & Richards, F.M. 1987 Tertiary templates for proteins. *J. molec. Biol* **193**, 775–791.

Ptitsyn, O.B. & Finkelstein, A.V. 1980 Similarities of protein topologies: evolutionary divergence, functional convergence or principles of folding. *Q. Rev. Biophys.* **13**, 339–386.

Richards, F.M. 1977 Areas, volumes, packing and protein structures. *A. Rev. biophys. Bioeng.* **6**, 151–176.

Ralston, E. & DeCoen, J.-L. 1974 Folding of polypeptide chains induced by the amino acid side-chains. *J. molec. Biol.* **83**, 393–420.

Reid, K.S.C., Lindley, P.F. & Thornton, J.M. 1985 Sulphur aromatic interactions in Proteins. *FEBS Lett.* **190**, 209–213.

Singh, J. & Thornton, J.M. 1990 SIRIUS An automated method for the analysis of the preferres packing arrangements between protein groups. *J. molec. Biol.* **211**, 595–615.

Singh, J. & Thornton J.M. 1985 The Interaction between phenylanine rings in proteins. *FEBS Lett.* **191**, 1–6.

Taylor, W.R. & Orengo, C.A. 1989 Protein structure alignment. *J. molec. Biol.* **208**, 1–22.

Wilmot, C.M. & Thornton, J.M. 1990 β-turns and their distortions: a proposed new nomenclature. *Protein Eng.* **3**, 479–493.

Williams, M.A. Goodfellow J.M. & Thornton J.M. 1994 Buried waters and internal cavities in monomeric proteins. *Protein Sci.* **3**, 1224–1235.

Zimmerman, S.S. Pottle, M.S. Némethy, G. & Scheraga, H.A. 1977. Conformational analysis of the 20 naturally occurring amino acid residues using ECEPP. *Macromolecules* **10**, 1–9.

Design of two-stranded and three-stranded coiled-coil peptides

STEPHEN BETZ, ROBERT FAIRMAN*, KARYN O'NEIL, JAMES LEAR‡
AND WILLIAM DEGRADO

The DuPont Merck Pharmaceutical Company, PO Box 80328, Wilmington, Delaware 19880-0328, U.S.A.

SUMMARY

The structural features required for the formation of two- versus three-stranded coiled coils have been explored using *de novo* protein design. Peptides with leucine at the 'a' and 'd' positions of a coiled-coil (general sequence: Leu_a Xaa_b Xaa_c Leu_d Glu_e Xaa_f Lys_g) exist in a non-cooperative equilibrium between unstructured monomers and helical dimers and helical trimers. Substituting valine into each 'a' position produces peptides which still form trimers at high concentrations, whereas substitution of a single asparagine at the 'a' position of the third heptad yields a dimer.

During the course of this work, we also re-investigated a helical propensity scale derived using a series of coiled-coil peptides previously believed to exist in a monomer–dimer equilibrium (O'Neil & DeGrado 1990). Detailed analysis of the concentration dependence of ellipticity at 222 nm reveals that they exist in a non-cooperative monomer–dimer–trimer equilibrium. However, the concentration of trimer near the midpoint of the concentration-dependent transition is small, so the previously determined values of $\Delta\Delta G_\alpha$ using the approximate monomer–dimer scheme are indistinguishable from the values obtained employing the complete monomer–dimer–trimer equilibrium.

1. INTRODUCTION

The *de novo* design of globular proteins with pre-determined structures and functions is a challenging goal that critically tests our understanding of the determinants of protein folding (DeGrado *et al.* 1991; Betz *et al.* 1993). In recent years, a variety of different proteins and peptides have been prepared including four-helix bundles (Handel *et al.* 1993; Choma *et al.* 1994), β-sandwich proteins (Quinn *et al.* 1994; Yan & Erickson 1994), and mixed α-β proteins (Beauregard *et al.* 1991; Tanaka *et al.* 1994). A simpler, but nevertheless challenging problem is the design of peptides that assemble into α-helical coiled-coil conformations (Graddis *et al.* 1993; O'Shea *et al.* 1993; Monera *et al.* 1994; Myszka & Chaiken 1994; Zhou *et al.* 1994;). The coiled coil is a simple structure which exhibits most of the characteristics of native proteins, including the formation of secondary structure stabilized by hydrophobic and electrostatic interactions. In coiled coils, two or more α-helices wrap around one another with a slight left-handed helical twist, forming a superhelix. Coiled coils exhibit a periodic primary structure which repeats every seven residues. Thus the design of a coiled-coil protein reduces to the problem of designing seven-residue peptide repeats which will lead to the formation of a given aggregation state.

In the past, the general structures and forces stabilizing coiled coils have been extensively studied (Crick 1953; Cohen & Parry 1990). Figures 1 and 2*a* illustrate structures for two-stranded and three-stranded coiled coils. In both cases, amino acid residues traditionally referred to as positions 'a' and 'd' project inward towards the superhelical axis. These residues are generally apolar and serve to stabilize the structure through hydrophobic interactions. The structures of two different classes of synthetic coiled coils have been determined. The coiled coil from GCN4 has been determined at high resolution, and found to form a two-stranded coiled coil that is very similar to the idealized structure shown in figure 1*a* (O'Shea *et al.* 1991). The sequence of this peptide consists of about four heptads. It contains Leu at each 'd' position and Val at each 'a' position with the exception of a single Asn at the 'a' position of the third heptad. The effects of substituting other residues into the 'a' and 'd' positions of all four heptads have been investigated. In this system, peptides with Leu at 'a' and Ile at 'd' form tetramers (Harbury *et al.* 1993), and peptides with Ile at all 'a' and 'd' positions form trimers (Harbury *et al.* 1994). These differences in association state have been rationalized in terms of the distinct packing preferences for Leu versus the β-branched amino acids Ile and Val.

In our own laboratory, we focused on a *de novo* designed peptide, originally intended to form a two-stranded coiled coil as a model system for determining

---

* Current Address: Division of Macromolecular Structure, Bristol Myers Squibb, PO Box 4000, Princeton, New Jersey 08543–4000, U.S.A.

‡ Current Address: Department of Biochemistry and Biophysics, School of Medicine, University of Pennsylvania, Philadelphia, Pennsylvania 19104–6059, U.S.A.

Phil. Trans. R. Soc. Lond. B (1995) **348**, 81–88
Printed in Great Britain

81

(a)

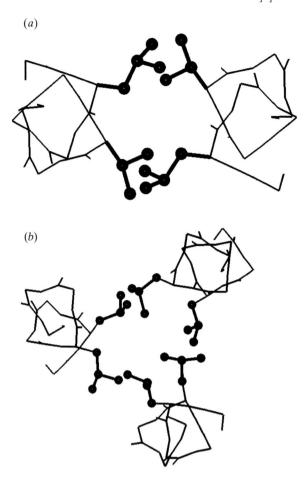

(b)

Figure 1. Hydrophobic core packing in coiled coils. 'a' and 'd' position side chains are shown as balls and sticks: (a) Parallel two-stranded coiled-coil from GCN4 (O'Shea et al. 1991); (b) Anti-parallel three-stranded coiled-coil from coil-Ser (Lovejoy et al. 1993).

helix propensities (O'Neil & DeGrado 1990). In an effort to determine an α-helical propensity scale, we wished to develop a system in which a peptide switches from being in the random-coil state to a fully α-helical state in a single transition that is thermodynamically linked to a two-state process, dimerization. A two-stranded coiled coil was designed that included a site for substituting 'guest' amino acids on the solvent-exposed face of the helix (see figure 2). The design of the helical dimer was based on the peptides of Hodges and co-workers (Lau et al. 1984). Following Hodges: the peptides contain Leu residues at the 'a' and 'd' positions, included to hydrophobically stabilize the structure, as well as interhelical electrostatic interactions between Glu and Lys residues on neighboring helices. The guest site is surrounded by neutral Ala residues to minimize the interactions between the side chain of the guest and those of the host. A set of 20 peptides was prepared with each of the commonly occurring amino acids substituted into the guest position. These peptides showed concentration-dependent circular dichroism (CD) spectra which could be analysed in terms of an equilibrium between random-coil monomers and α-helical dimers. The differences between the free energies of dimerization for the various peptides ($\Delta\Delta G_\alpha$) were dependent on the helix-forming character of the guest amino acid, which

allowed the derivation of a new scale of intrinsic helical propensities (O'Neil & DeGrado 1990).

After completing determination of the helix propensities, the crystal structure of one of the peptides, coil-Ser (named for the amino acid at the guest position) was determined and found to be a trimer rather than the expected dimer (Lovejoy et al. 1993). As expected, the structure consists of a pair of parallel helices, with the 'host' site at the 'f' position fully exposed to solvent. However, a third helix docks against the dimer anti-parallel to the other pair (see figures 1 b and 2 a). Sedimentation equilibrium showed that the peptide formed trimers in solution at micromolar concentrations (Lovejoy et al. 1993). These findings stimulated a thorough re-examination of the thermodynamics of self-association of the Coil-Xaa family of peptides (see figure 2 b). Here we show that the peptides exist in a non-cooperative monomer–dimer–trimer equilibrium in which the free energy of dimerization of two monomers is approximately equal to the free energy of adding a monomer to a dimer to form a trimer. This analysis allowed the calculation of a corrected helix thermodynamic scale ($\Delta\Delta G_\alpha$). Fortunately, the values of $\Delta\Delta G_\alpha$ are indistinguishable from the earlier values within the error of our measurements.

We have also investigated the structural features that cause the peptide to form trimers rather than dimers. Figure 2 illustrates the sequence coil-Ser as compared to the two-stranded coiled-coil of GCN4. Both have Leu residues at the 'd' position, but differ in the residues at the 'a' position, including the presence of Trp in the first heptad of Coil-Ser. This residue was included as a spectroscopic probe, and could conceivably direct the antiparallel orientation of the helices in the trimer by disfavouring an all-parallel topology. Models suggest that if the peptides adopt a parallel trimer, the bulky Trp residues bump into one another in an unfavourable interaction that is absent in the antiparallel structure. We therefore synthesized the peptide coil-$L_a L_d$, in which the Trp is changed to Leu resulting in a peptide with Leu in every 'a' and 'd' position. GCN4 contains Val residues at each of its 'a' positions so we also investigated changing all the Leu residues in the 'a' positions to Val in peptide coil-V_a-L_d. Additionally, we investigated the role of the Asn at an 'a' position in GCN4 through the synthesis of Asn_{16}-coil-V_a-L_d. Finally, in an effort to reconcile our results with earlier results by Hodges on his coiled-coil peptides, we prepared TM-43, that also contains Leu residues at positions 'a' and 'd'.

2. METHODS
(a) Peptides

The peptides TM-43, Asn_{16}-coil-V_a-L_d, coil-V_a-L_d, and coil-L_a-L_d were prepared using Fmoc-protected amino acids and techniques described previously (Choma et al. 1994).

(b) Sedimentation equilibrium

Sedimentation equilibrium analysis was performed using a Beckman XLA analytical ultracentrifuge. The

(a)

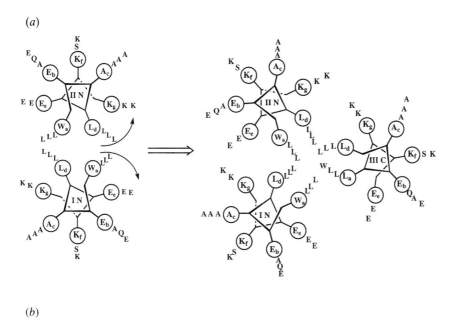

(b)

peptide			heptad position			
	g	abcdefg	abcdefg	abcdefg	abcdefg	abcdefg
coil-Xaa	E	WEALEKK	LAALE**X**K	LQALEKK	LEALEHG	
coil-L_aL_d	E	LEALEKK	LAALESK	LQALEKK	LEALEHG	
coil-V_aL_d	E	VEALEKK	VAALESK	VQALEKK	VEALEHG	
Asn_{16}-coil-V_aL_d	E	VEALEKK	VAALESK	NQALEKK	VEALEHG	
TM-43	K	LEALEGK	LEALEGK	LEALEGK	LEALEGK	(LEALEGK)$_2$
GCN4	R	MKQLEDK	VEELLSK	NYHLENE	VARLKKL	VGER

Figure 2. (a) Helical wheel representations of the desired packing of a parallel two-stranded coiled-coil versus the observed anti-parallel three-stranded coiled-coil in coil-Ser. The structures are related by the conceptual rotation of the two parallel helices (indicated by arrows), followed by the packing of the third helix in an antiparallel orientation. (b) Amino acid sequences of the peptides discussed in this paper, the top row indicates position within the coiled-coil heptad.

conditions for measuring the sedimentation behaviour, and analysis of coil-Ser, coil-Ala, coil-Arg, coil-Trp, Asn_{16}-coil-V_a-L_d, coil-V_a-L_d, and coil-L_a-L_d have been published previously (Lovejoy *et al.* 1993). Initial peptide concentrations were 20 μM and 200 μM. The conditions for TM-43 were chosen to match those of Hodges and co-workers (Lau *et al.* 1984): 50 mm potassium phosphate, pH 7.0 or 2.5, containing either 0.1 M KCl or 1.1 M KCl. The samples were centrifuged at 25000, 40000, and 48000 rpm; equilibrium was determined when successive radial scans at the same speed were indistinguishable. Partial specific volumes were determined by the residue-weighted average method of Cohn and Edsall (Cohn & Edsall 1943; Harding *et al.* 1992), as well as the H_2O–D_2O difference method (Edelstein & Schachman 1973) and found to agree within 0.01 ml g^{-1}. Solution densities were estimated using solute concentration-dependent density tables in the CRC Handbook of Chemistry and Physics. The aggregation state was determined by

monomer-$n$mer equilibria. Also, fitting the data as a single species yielded molecular masses consistent with these aggregation states. Data were analysed with the software package Igor Pro (WaveMetrics, Inc.).

(c) *Analysis of the concentration dependence of the CD spectra*

The data can be fit to an empirically derived Hill coefficient, which examines the cooperativity of the association according to equations (1) and (2):

$$K_H = [P_{mon}]^n/[P_n], \tag{1}$$

$$[P_T] = [P_{mon}]^n/(K_H + [P_{mon}]), \tag{2}$$

in which P_{mon}, P_n, and P_T are the concentrations of the monomer, $n$mer and total peptide, respectively, n is the Hill coefficient, and K_H is a dissociation constant. Equation (2) is solved numerically for $[P_{mon}]$ using the root function of Mlab (Civilized Software), and inserted into equation (3):

$$[\theta_{222}] = [\theta_{nmer}](f) = [\theta_{nmer}](([P_{mon}]^n/K_H)/P_T, \quad (3)$$

in which $[\theta_{nmer}]$ is the mean residue ellipticity of the associated form of the peptide and f is the fraction of the peptide in the associated form. The parameters, $[\theta_{nmer}]$, n, and K_H are then determined for each peptide with the nonlinear least squares method using Mlab. This method of analysis assumes that the ellipticity of the monomer is zero; and was confirmed by allowing the ellipticity of the monomer to be an additional adjustable parameter, in which case the determined value was consistently within 3000 deg cm² dmol⁻¹ of zero.

Alternatively the data could fit a monomer–dimer–trimer equilibrium as described in equation (7) in Results. The mass balance equation is:

$$[P_T] = M + 2[D] + 3[T]$$
$$= [M] + 2[M]^2/K_1 + 3[M]^3/K_1K_2. \quad (4)$$

The mean residue ellipticity at a given peptide concentration is given by:

$$[\theta_{222}] = ([M][\theta_M] + 2[D][\theta_D] + 3[T][\theta_T])/[P_T]. \quad (5)$$

in which the mean residue ellipticity of the dimer, $[\theta_D]$, and the trimer, $[\theta_D]$, are assumed to be identical; and the ellipticity of the monomer, $[\theta_m]$, is assumed to be zero. The parameters $[\theta_D]$, $[\theta_T]$, $[K_1]$ and $[K_2]$ are then determined for each peptide by the nonlinear least squares method using Mlab. In several calculations, the ellipticity of the monomer was included as an additional adjustable parameter; the determined value of this parameter in each case was consistently within 3000 deg cm² dmol⁻¹ of zero.

The data were also analysed according to equations (8) and (9) in Results. With this approach a given peptide is chosen as a reference peptide, and values of K_1 and K_2 are calculated from the concentration dependence of its CD spectrum. The data for each additional peptide are then calculated using the mass balance equation:

$$[P_T] = [M] + 2[D] + 3[T]$$
$$= [M] + 2[M]^2/K_1' + 3[M]^3/K_1'K_2', \quad (6)$$

in which K_1' and K_2' are as defined in equations (8) and (9). Equation (6) is once again numerically solved for [M] and inserted into equation (5). Now, however, because the values of K_1 and K_2 are known, the only fitted parameters are $\Delta\Delta G_\alpha$ and the ellipticities of the dimers and trimers (assumed to be equal). Coil-Ser was chosen because we had an extensive set of data for this peptide with values of K_1 and K_2 of 8.1 ± 2.5 and 5.0 ± 1.3, respectively. The values of $\Delta\Delta G_\alpha$ were then calculated for each of the remaining peptides and were referenced to Gly.

3. RESULTS

(a) Sedimentation equilibrium studies

We had adequate quantities of four peptides previously used in the helix propensity experiments (O'Neil &

DeGrado 1990) to determine their aggregation states by sedimentation equilibrium (see table 1). At neutral pH, room temperature, with no denaturants and peptide concentrations ranging from 20–200 μM, the peptides sediment as single, homogeneous species with molecular masses consistent with a trimeric aggregation state. These data are consistent with earlier studies that showed the peptides self-assembled into highly stable helical aggregates that only dissociate in the presence of 4–7 M urea (O'Neil & DeGrado 1990).

The peptides coil-V_a-L_d and coil-L_a-L_d also were also found to be trimeric under these conditions, and only the peptide Asn_{16}-coil-V_a-L_d was found to form dimers. At an initial loading concentration of 1 mM this peptide sedimented as a single species with a molecular mass close to that expected for a dimer (see table 1).

The trimeric association state observed for coil-L_a-L_d is in marked contrast with the dimeric state reported by Hodges and co-workers for the parent coiled-coil peptide (Lau *et al.* 1984). We therefore prepared the parent coiled-coil peptide, TM-43, and determined its aggregation state by sedimentation equilibrium (see figure 3) at initial loading concentrations of 60 μM or 600 μM, pH 7.0 or 2.5, and 0.1 M KCl or 1.1 M KCl. At pH 7.0 (either combination of peptide and salt concentration), the data clearly indicate that this peptide forms trimers (see table 2). At pH 2.5, the peptide forms higher-order aggregates that are likely tetramers. One potential explanation for the difference between these findings and earlier data reported by Hodges (Lau *et al.* 1984; Zhu *et al.* 1993) might be in the values used for the partial specific volume and density. Using the H_2O–D_2O difference method (Edelstein & Schachman 1973), we determined the partial specific

Table 1. *Equilibrium sedimentation of coil-Xaa peptides*

(Conditions: 10 mM MOPS, pH 7.5. The molecular masses were determined using a partial specific volume calculated by the residue weighted average method (Cohn & Edsall 1943; Harding *et al.* 1992) as well as the H_2O–D_2O difference method (Edelstein & Schachman 1973) and found to agree within 0.01 ml g⁻¹. The partial specific volume for each of these peptides is 0.76 ml g⁻¹, except as noted. Solution densities were estimated by solute concentration-dependent density tables in the CRC Handbook of Chemistry and Physics. We estimate that the uncertainty in these measurements arises chiefly from the uncertainty in the partial specific volume, leading to an uncertainty of 10–20% in the observed molecular masses.)

peptide	monomeric MM	observed MM
coil-Ala	3331	10 500
coil-Arg	3416	10 800
coil-Trp	3446	10 800
coil-Ser[a]	3347	10 258
coil-$V_a L_d$	3218	10 200
coil-$L_a L_d$	3274	9 700
Asn_{16}-coil-$v_a L_d$	3233	7 100[b]

[a] Data are from Lovejoy *et al.* (1993).
[b] Mean value of runs at rotor speeds of 20, 30, and 40 krpm. The partial specific volume was calculated to be 0.75 ml g⁻¹.

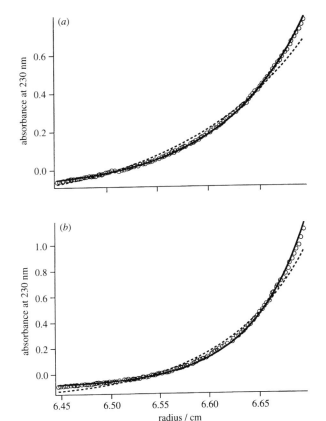

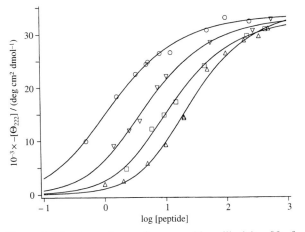

Figure 4. The dependence of mean residue ellipticity, $[\theta_{222}]$, on concentration for coiled-coil peptides. The raw data are shown as symbols: (circles) coil-Ala, (inverted triangles) coil-Ser, (squares) coil-Asp, (triangles) coil-D-Ala. The calculated fits are shown as solid lines. The Hill coefficients are presented in table 3.

Figure 3. Sedimentation equilibrium of the TM-43 peptide. Circles represent the raw data, a solid line shows the theoretical fit to trimer, and a dashed line shows the theoretical fit to dimer. The data in the top panel were acquired at 35000 rpm. The data in the bottom panel were acquired at 48000 rpm. The conditions were 60 μM TM-43, 1.1 MKCl, pH 7.0.

Table 2. *Equilibrium sedimentation of TM-43 peptide*

(Samples were run at initial loading concentrations of 60 μM and 600 μM in 50 mM potassium phosphate, except as noted. The molecular mass of the monomer is 4632. The prevalent species was determined by monomer-$n$mer equilibria. Molecular mass calculated assuming only one species in solution. The partial specific volume was calculated by the H_2O–D_2O difference method (Edelstein & Schachman 1973) to be 0.76 ml g^{-1} in 50 mM potassium phosphate, 1.1 M KCl, and 0.75 ml g^{-1} in 10 mM MOPS.)

pH	[KCl] M	prevalent species	single-species molecular mass	association state
7.0	1.1	trimer	13700	3.0
7.0	0.1	trimer	14200	3.1
2.5	1.1	tetramer	19100	4.1
2.5	0.1	tetramer	18000	3.9
7.5[a]	0.0	trimer	13400	2.9
7.5[a]	0.1	trimer	14600	3.1

[a] 10 mM MOPS buffer.

volume of TM-43 to be 0.76 ml g^{-1} under the conditions used here. The value used by Lau *et al.* (1984) was estimated to be 0.70 ml g^{-1}; the employment of this low value will lead to an underestimation of the aggregate's molecular mass. Solution

densities used by Hodges and co-workers (Lau *et al.* 1984; Zhu *et al.* 1993) were not reported.

(b) *Concentration dependence of the CD spectra of the original coiled-coil peptides*

In our earlier studies, self association of the peptides was measured in the presence of 5.0 M urea, where the oligomeric forms of these peptides are considerably less stable. Under these conditions, the concentration-dependent equilibrium between fully non-helical monomers and helical oligomers can be monitored readily by measuring the concentration dependence of their CD spectra. Figure 4 illustrates the concentration dependence of the helical content of several coiled-coil peptides as monitored by their mean residue ellipticity at 222 nm $[\theta_{222}]$. These curves can be analysed by a monomer–dimer–trimer equilibrium:

$$T \overset{K_2}{\rightleftharpoons} D + M \overset{K_1}{\rightleftharpoons} 3M,$$

$$K_1 = [M]^2/[D]; \quad K_2 = [D][M]/[T]. \quad (7)$$

Two possible factors limit this equilibrium scheme: if K_1 is much smaller than K_2, then monomers and dimers would be the only significantly populated species in the transition zone and trimers would be formed only at considerably higher concentration. In this case, the Hill coefficient for the process would be 2.0, and the curves would be indistinguishable from simple monomer–dimer equilibria. Conversely, if K_2 is much smaller than K_1, then the concentration of dimers would be very small (relative to monomers and trimers) in the transition zone. In this case the Hill coefficient would be 3.0 and the data could be analysed as a cooperative monomer–trimer equilibrium. If the Hill coefficient for the process falls between these two limiting values, then K_1 and K_2 are reasonably close in magnitude and can be determined through careful analysis of the curve.

The Hill coefficients (n) for the 20 previously

Table 3. *Helix formation parameters and Hill coefficients for coiled-coil peptides*

(The values of $\Delta\Delta G_\alpha$ and n were determined using previously published data (O'Neil & DeGrado 1990; Fairman *et al.* 1992) and for the charged residues, have been corrected for electrostatic effects by extrapolating the value of $\Delta\Delta G_\alpha$ to 1.0 M NaCl.)

amino acid	Hill Coefficient	$\Delta\Delta G_\alpha$ (old) kcal mol^{-1}	$\Delta\Delta G_\alpha$ (new kcal mol^{-1}
Ala	2.25 (0.15)	−0.77	−0.71
D-Ala	2.42 (0.21)	0.18	0.34
Arg	2.34 (0.34)	−0.68	−0.70
Asp	2.37 (0.15)	−0.15	−0.10
Asn	2.19 (0.09)	−0.07	−0.01
Cys	2.51 (0.35)	−0.23	−0.22
Gln	2.99 (0.42)	−0.33	−0.33
Glu	2.40 (0.14)	−0.27	−0.21
Gly	2.90 (0.26)	0.00	0.00
His	2.22 (0.12)	−0.06	0.03
Ile	2.36 (0.17)	−0.23	−0.17
Leu	3.15 (0.63)	−0.62	−0.52
Lys	2.20 (0.26)	−0.65	−0.58
Met	2.46 (0.27)	−0.50	−0.42
Phe	1.90 (0.18)	−0.41	−0.37
Ser	2.31 (0.22)	−0.35	−0.27
Thr	2.23 (0.21)	−0.11	−0.09
Tyr	2.82 (0.15)	−0.17	−0.06
Trp	1.93 (0.68)	−0.45	−0.45
Val	2.55 (0.21)	−0.14	−0.16

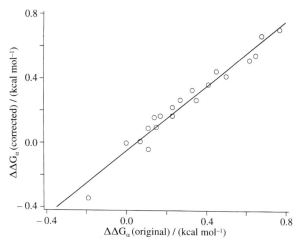

Figure 5. Comparison of revised and original values of $\Delta\Delta G_\alpha$. The individual data points are shown as circles, the linear regression is shown as a solid line.

reported coiled-coil peptides were each determined as described in the Methods (see table 3) from our published data and found to be 2.41 ± 0.07 (standard error of the mean). Two peptides were found to have Hill coefficients as low as 1.9 (coil-Phe and coil-Trp) and four had Hill coefficients of 2.8–3.15 (coil-Tyr, coil-Leu, coil-Gly and coil-Gln). However, the values of n for these peptides had larger relative uncertainties (see table 3), suggesting that deviations from the mean arose from errors in the data sets for these peptides rather than actual differences in the cooperativity. To test this idea further, we selected one peptide – coil-Tyr – for further study. A triplicate data set gave a Hill coefficient of 2.3 ± 0.1, suggesting that the outliers represented error in our previous data rather than actual differences in the cooperativity of association. This is structurally reasonable because the helix–helix interface should be very similar for each of the peptides as the changes in their sequences occur on their solvent-exposed faces.

A Hill coefficient of 2.4 indicates that the values of K_1 and K_2 are of similar magnitude. This was confirmed by fitting the individual data sets for the peptides to a monomer–dimer–trimer equilibrium as described in Methods. The average value of K_2/K_1 for the entire set of peptides was 0.6 corresponding to an energetic difference of 0.3 ± 0.15 kcal mol^{-1} (standard error of the mean).

To obtain estimates of the helix propensities of the amino acids from the monomer–dimer–trimer equilibria it is useful to consider how a change in the guest residue would affect the equilibrium constants, K_1 and

K_2. Consider a given peptide with a given K_1 and K_2. Now another peptide with a different guest residue will have a monomer–dimer equilibrium constant, K_1, which is related to K_1 for the original peptide by equation (8):

$$K_1' = K_1\, e^{-(2\Delta\Delta G\alpha/RT)}, \tag{8}$$

in which $\Delta\Delta G_\alpha$ is the difference in the free energy of dissociation between the two peptides (on a 'per monomer' basis), and the factor of two is included because two monomers join to form a dimer. After our earlier treatment we associate this free energy difference ($\Delta\Delta G_\alpha$) primarily with the difference in the propensity of helix formation for the two amino acids because the mutation occurs on the solvent-exposed face of the helix at a site distant from the helix–helix interface. Further, the dissociation constant for adding a single random-coil monomer to the helical dimer in the second step is given in equation (9) by analogy to equation (8)

$$K_2' = K_2\, e^{-(\Delta\Delta G\alpha/RT)}. \tag{9}$$

However, in this step only one peptide switches from a random-coil monomer to a helical dimer so the value of $\Delta\Delta G_\alpha$ is not multiplied by two. We assume that $\Delta\Delta G_\alpha$ is identical for each monomer in the trimer because the local environment of the host sites are extremely similar as judged by the crystal structure. Using this treatment we obtained $\Delta\Delta G_\alpha$ for each of the amino acids referenced to Gly.

Table 3 provides the values of $\Delta\Delta G_\alpha$ for the commonly occurring amino acids derived from the simple monomer–dimer equilibria versus the more complete monomer–dimer–trimer equilibria (a plot is displayed in figure 5). The data are linearly related by the equation [$\Delta\Delta G_\alpha$ (new) $= 1.006 \times \Delta\Delta G_a$ (old) -0.05; $r^2 = 0.98$]. Within our experimental error (0.1 kcal mol^{-1}) there are no significant differences between the two data sets.

4. DISCUSSION

The data in this paper clearly demonstrate that the peptides in this series, which contain Leu at positions 'a' and 'd', Glu at position 'e' and Lys at position 'g' of the coiled coil, exist in a non-cooperative monomer–dimer–trimer equilibrium. The relative amounts of the monomeric, dimeric and trimeric species as a function of the total peptide concentration for coil-Ala are shown in figure 6. We also explored the features that cause the peptides to form trimers rather than the originally designed dimers. Replacing Leu with Val at each 'a' position led to peptides that still formed trimers, and preliminary measurements of the concentration dependence of this peptide's CD signal indicate that the association is more cooperative; the addition of a monomer to a preformed dimer to form a trimer is more favourable than the initial dimerization of the two monomers. In contrast, addition of a single Asn residue at an 'a' position caused the peptide to form only dimers. This result parallels results previously described by Kim, Alber and their coworkers for the coiled-coil of GCN4 (O'Shea *et al.* 1991, 1993). The structural basis for this result can be appreciated by considering the effect of burying an Asn at the helix–helix interface of a dimer versus a trimer. In a dimer, the carboxamide side-chain from neighboring Asn residues can form a hydrogen bond between the carbonyl of one residue and the -NH$_2$ of the other (O'Shea *et al.* 1991). More importantly, having made these stabilizing interactions, the remaining portions of this polar side-chain can be hydrated through interactions with water. Model building suggests that in a parallel trimer, the same side chain–side chain hydrogen bonding can be realized. However, the Asn is fully buried in the interior of the structure without forming hydrogen bonds to all its polar functions. If the peptide forms an antiparallel trimer, as in coil-Ser, two Asn residues originating from the parallel pair of helices can form stabilizing H-bonded interactions, but the third Asn would be forced to be buried among Leu

residues. Therefore the dimer is stabilized over the trimer by hydration effects when a single polar residue is introduced into the sequence. Thus, this change provides specificity to the structure, albeit at the price of thermodynamic stability. (Our preliminary data indicate that the addition of Asn destabilizes the free energy of dimerization by several kcal mol⁻¹.)

From the perspective of *de novo* protein design, the non-specificity of the aggregation of coil-Ser was far from optimal. In a designed protein one seeks to obtain a structure that has a unique conformational state or association state; unlike the non-cooperative assembly seen for these peptides. If one wishes to stabilize a parallel dimer relative to a trimer, the addition of an Asn residue at an 'a' position appears to be a successful strategy. Alternatively, the addition of a Cys residue to form a disulphide-bonded dimer has been reported by Hodges to stabilize the dimer, although some higher-order oligomers were also formed (Zhou *et al.* 1993). The addition of β-branched amino acids, however, may stabilize the trimeric state although we have not yet investigated this in detail. Perhaps the most straightforward method to stabilize an antiparallel trimer is to entropically stabilize the three-helix bundle state through the incorporation of links between the helices. Work along these lines is showing considerable progress in our labs.

The primary motivation of our original design of two-stranded coiled coils was to design a system for exploring the helix propensities of the 20 commonly occurring amino acids. The finding that the peptides we had designed formed trimers as well as dimers, necessitated a thorough re-examination of our data. This analysis shows that the previously reported values of $\Delta\Delta G_\alpha$ obtained using a simple monomer-dimer treatment were, allowing for experimental error, indistinguishable from the values obtained in this work using a monomer–dimer–trimer treatment. This is because the values of $\Delta\Delta G_\alpha$ are obtained at the midpoint of the transition, where the amount of peptide in the trimeric state is about 15 % of the total (see figure 6). However, explicit treatment of the effect of trimer formation does remove a small (~ 0.05 kcal mol⁻¹), positive systematic error in the $\Delta\Delta G_\alpha$ values. Also, it is worth noting that although the new $\Delta\Delta G_\alpha$ values result in a different rank ordering of the helix propensities of the various amino acids, the absolute differences responsible for these rank changes are well within experimental uncertainty and, in fact, the correlation with other scales for helix formation (Park *et al.* 1993; Chakrabartty *et al.* 1994; Munoz & Serrano 1994) remains excellent and should continue to be a valuable aid in protein design.

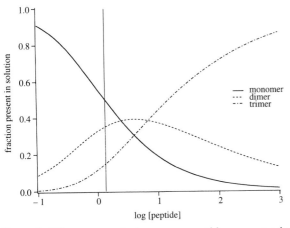

Figure 6. Oligomer population versus peptide concentration (µM) for coil-Ala. The curves are calculated from the equilibrium constants of the concentration dependent CD data. A solid line represents monomer, a dashed line represents dimer, and a dot-dashed line represents trimer. The vertical line represents the midpoint concentration of the association curve of coil-Ala.

REFERENCES

Beauregard, M., Goraj, K., Goffin, V., Heremans, K., Goormaghtigh, E., Ruysschaert, J. & Martial, J.A. 1991 Spectroscopic investigation of structure in octarellin (a de novo protein designed to adopt the α/β-barrel packing). *Protein Engng.* **4**, 745–749.

Betz, S.F., Raleigh, D.P. & DeGrado, W.F. 1993 De novo

protein design: from molten globules to native-like states. *Curr. Opin. struct. Biol.* **3**, 601–610.

Chakrabartty, A., Kortemme, T. & Baldwin, R.L. 1994 Helix Propensities of the Amino Acids Measured in Alanine-Based Peptides Without Helix-Stabilizing Side-Chain Interactions. *Protein Sci.* **3**, 843–852.

Choma, C.T., Lear, J.D., Nelson, M.J., Dutton, P.L., Robertson, D.E. & DeGrado, W.F. 1994 Design of a heme-binding four-helix bundle. *J. Am. chem. Soc.* **116**, 856–865.

Cohen, C. & Parry, D.A.D. 1990 α-helical coiled coils and bundles: how to design an α-helical protein. *Proteins* **7**, 1–15.

Cohn, E.J. & Edsall, J.T. 1943 *Proteins, amino acids and peptides as ions and dipolar ions*, pp. 370–377. New York: Reinhold Publishing Corp.

Crick, F.H.C. 1953 The Fourier Transform of a Coiled Coil. *Acta Crystallogr.* **6**, 685–689.

DeGrado, W.F., Raleigh, D.P. & Handel, T. 1991 De novo Protein Design: What Are We Learning? *Curr. Opin. struct. Biol.* **1**, 984–993.

Edelstein, S.J. & Schachman, H.K. 1973 Measurement of partial specific volume by sedimentation equilibrium in H_2O–D_2O solutions. *Meth. Enzymol.* **27**, 82–98.

Fairman, R., Anthony-Cahill, S.J. & DeGrado, W.F. 1992 The Helix-Forming Propensity of D-Alanine in a Right-Handed α-Helix. *J. Am. chem. Soc.* **114**, 5458–5459.

Graddis, T.J., Myszka, D.G. & Chaiken, I.M. 1993 Controlled formation of Model Homo- and Heterodimer Coiled-coil Polypeptides. *Biochemistry* **32**, 12664–12671.

Handel, T.M., Williams, S.A. & DeGrado, W.F. 1993 Metal ion-dependent modulation of the dynamics of a designed protein. *Science, Wash.* **261**, 879–885.

Harbury, P.B., Zhang, T., Kim, P.S. & Alber, T. 1993 A Switch Between Two-, Three-, and Four- Stranded Coils in GCN4 Leucine Zipper Mutants. *Science, Wash.* **262**, 1401–1407.

Harbury, P.B., Kim, P.S. & Alber, T. 1994 Crystal structure of an isoleucine-zipper trimer. *Nature, Lond.* **371**, 80–83.

Harding, S.E., Rowe, A.J. & Horton, J.C. 1992 *Analytical ultracentrifugation in biochemistry and polymer science.* Cambridge: The Royal Society of Chemistry.

Lau, S.Y.M., Taneja, A.K. & Hodges, R.S. 1984 Synthesis of a Model Protein of Defined Secondary and Quaternary Structure. *J. biol. Chem.* **259**, 13253–13261.

Monera, O.D., Kay, C.M. & Hodges, R.S. 1994 Electrostatic Interactions Control the Parallel and Antiparallel Orientation of α-Helical Chains in Two-Stranded α-Helical Coiled Coils. *Biochemistry* **33**, 3862–3871.

Munoz, V. & Serrano, L. 1994 Elucidating the folding problem of helical peptides using empirical parameters. *Nature struct. Biol.* **1**, 399–409.

Myszka, D.G. & Chaiken, I.M. 1994 Design and Characterization of an Intramolecular Antiparallel Coiled-coil Peptide. *Biochemistry* **33**, 2363–2372.

O'Neil, K.T. & DeGrado, W.F. 1990 A Thermodynamic Scale for the Helix-Forming Tendencies of the Commonly Occurring Amino Acids. *Science, Wash.* **250**, 646–651.

O'Shea, E.K., Klemm, J.D., Kim, P.S. & Alber, T.A. 1991 X-ray structure of the GCN4 leucine zipper, a two-stranded, parallel coiled coil. *Science, Wash.* **254**, 539–544.

O'Shea, E.K., Lumb, K.J. & Kim, P.S. 1993 Peptide Velco: Design of a Heteromeric Coiled Coil. *Curr. Biol.* **3**, 658–667.

Park, S.H., Shalongo, W. & Stellwagen, E. 1993 Residue helix parameters obtained from dichroic analysis of peptides of defined structure. *Biochemistry* **32**, 7048–7053.

Quinn, T.P., Tweedy, N.B., Williams, R.W., Richardson, J. S. & Richardson, D.C. 1994 Betadoublet: *De novo* Design, Synthesis, and Characterization of a β-sandwich Protein. *Proc. natn. Acad. Sci. U.S.A.* **91**, 8747–8751.

Tanaka, T., Kuroda, Y., Kimura, H., Kidokoro, S. & Nakamura, H. 1994 Cooperative deformation of a de novo designed protein. *Prot. Engng.* **7**, 969–976.

Yan, Y. & Erickson, B.W. 1994 Engineering of Betabellin 14D: Disulfide-induced Folding of a β-sheet Protein. *Protein Sci.* **3**, 1069–1073.

Zhou, N.E., Kay, C.M. & Hodges, R.S. 1993 Disulfide bond contribution to protein stability: Positional effects of substitution in the hydrophobic core of the two-stranded α-helical coiled coil. *Biochemistry* **32**, 3178–3187.

Zhou, N.E., Kay, C.M. & Hodges, R.S. 1994 The Role of Interhelical Ionic Interactions in Controlling Protein Folding and Stability. *J. molec. Biol.* **237**, 500–512.

Zhu, B.-Y., Zhou, N.E., Kay, C.M. & Hodges, R.S. 1993 Packing and hydrophobicity effects on protein folding and stability: Effects of beta-branched amino acids. valine and isoleucine, on the formation and stability of two stranded alpha-helical coiled coils/leucine zippers. *Protein Sci.* **2**, 383–394.

Nascent chains: folding and chaperone interaction during elongation on ribosomes

KOSTAS TOKATLIDIS[1], BERTRAND FRIGUET[1],
DOMINIQUE DEVILLE-BONNE[1], FRANÇOISE BALEUX[2],
ALEXEY N. FEDOROV[3], AMIEL NAVON[4],
LISA DJAVADI-OHANIANCE[1] AND MICHEL E. GOLDBERG[1]

[1] *Unité de Biochimie Cellulaire, Institut Pasteur, 28 Rue du Dr Roux, 75015 Paris, France*
[2] *Unité de Chimie Organique, Institut Pasteur, 28 Rue du Dr Roux, 75015 Paris, France*
[3] *Institute of Protein Research, Russian Academy of Sciences, 142292 Puschino, Moscow Region, Russia*
[4] *Department of Life Science, Bar Ilan University, 52900 Ramat Gan, Israel*

SUMMARY

Monoclonal antibodies that detect folding intermediates *in vitro* were used to monitor the appearance of folded polypeptide chains during their synthesis on the ribosomes. Nascent immunoreactive chains of the bacteriophage P22 tail-spike protein and of the *Escherichia coli* β_2 subunit of tryptophan-synthase were thus identified, suggesting that they can fold on the ribosomes. Moreover, the immunoreactivity of ribosome-bound tryptophan-synthase β-chains of intermediate lengths was shown to appear with no detectable delay compared to their synthesis. This suggested that β-chains start folding during their elongation on the ribosomes.

However, newly synthesized incomplete β-chains were shown to interact with chaperones while still bound to the ribosome. Because of the peculiar properties of the epitope recognized by the anti-tryptophan-synthase monoclonal antibody used, it could not be concluded whether the immunoreactivity of the nascent β-chains resulted from their ability to fold cotranslationally or from their association with chaperones which might maintain them in an unfolded, immunoreactive state.

1. INTRODUCTION

Our current understanding of the mechanisms by which a protein acquires its native conformation is based on *in vitro* studies of the folding of pure, complete polypeptide chains. *In vivo*, the polypeptide chain grows progressively, starting at its N-terminal end (attached to the ribosome) which is in contact with the complex intracellular medium. Whether or not the observations made *in vitro* are relevant to the process that occurs *in vivo* is still poorly documented, essentially because of the considerable difficulties encountered when characterizing the conformation of nascent chains. This problem is further complicated by the fact that these chains can only be produced in very small quantities and in the presence of numerous other cellular components. It seemed possible to us that these difficulties could be overcome by taking advantage of the remarkable specificity of antibodies and the high sensitivity of immunodetection methods. We therefore used monoclonal antibodies (MAbs) to characterize the conformation of ribosome-bound nascent chains from the bacteriophage P22 tailspike protein and of the β_2 subunit of *E. coli* tryptophane synthase.

Each of these antibodies was shown, through *in vitro* refolding experiments, to recognize an epitope carried by the corresponding polypeptide chains only if they have undergone at least some folding steps (Murry-Brelier & Goldberg 1988; Blond-Elguindi & Goldberg 1990; Friguet *et al.* 1994). Attempts were made to use these antibodies for detecting the presence of folded, ribosome-bound nascent chains during biosynthesis of P22 tailspike protein and of tryptophan-synthase β-chains. Here we describe the results that were obtained by using this approach, the experimental difficulties that were overcome and the uncertainties that remain in their interpretation.

2. EVIDENCE FOR THE EXISTENCE OF FOLDED RIBOSOME-BOUND P22 TAILSPIKE NASCENT CHAINS

Seckler *et al.* (1989) identified several intermediates during the *in vitro* folding of the guanidine unfolded tailspike protein of bacteriophage P22 and characterized the kinetics of their appearance. These intermediates closely resemble those identified during the *in vivo* maturation of the tailspike (Goldenberg & King 1982; Haase-Pettingell & King 1988). We prepared and characterized a panel of mAbs which recognize the native tailspike protein (Friguet *et al.* 1990) and used them to monitor the kinetics of regain of immunoreactivity during the refolding of the tailspike protein (Friguet *et al.* 1994). Three mAbs were thus found to

Phil. Trans. R. Soc. Lond. B (1995) **348**, 89–95
Printed in Great Britain

89

detect an antigenic determinant, different for each mAb, that reappears with kinetics corresponding to a specific folding step as identified by Seckler *et al.* (1989). The antigenic site recognized by mAb 236-3 is formed simultaneously with a folded monomer competent for protrimer formation; mAb 155-3 recognizes the protein from the moment it forms a heat labile protrimer; mAb 33-2 recognizes only the heat-stable native trimer (Friguet *et al.* 1994). Because these three antibodies could recognize three intermediates at different stages of the folding pathway, we used them to investigate whether such folded intermediates were present on ribosome-bound nascent chains. Cells infected with deficient phages and expressing high levels of the tailspike protein were lysed, the debris removed by centrifugation, and the ribosomal fraction obtained by centrifugation on a sucrose gradient. To test for the presence of ribosome-bound immunoreactive nascent chains, the ribosomal fraction was immunoprecipitated with each of the mAbs and assayed for the presence of ribosomal proteins in the immunoprecipitate: this was done by the following method.

The ribosomal fraction, prepared as outlined above, was incubated with the various mAbs and subjected to immunoadsorption on Sepharose beads coupled to Protein G. After low speed centrifugation to pellet the beads, the precipitate was washed several times. The immunoadsorbed proteins were dissolved in sodium dodecyl-sulphate (SDS) and submitted to SDS-polyacylamide gel electrophoresis (SDS-PAGE) and Western blotting. The blot was developed with a rabbit immune serum directed against the ribosomal protein L20 and with anti-rabbit immunoglobulins coupled to alkaline phosphatase. When mAb 236-3 was used, a band was clearly visible at a position corresponding to approximately 13.5 kDa, the molecular mass of the L20 protein. This band could not be detected when the anti-tailspike mAb 155-3 or when a control (nonspecific) mAb were used. This unambiguously showed that some ribosome-bound chains can indeed react with mAb 236-3, and are therefore likely to have undergone the folding steps that generate the epitope recognized by this mAb.

These experiments were only preliminary, however, and provided no information on the nature (in particular on the length and amount) of the immunoreactive chains. Nor did they provide any information on the conformation of the nascent chains as no attempt was made to investigate the affinity of the mAb for the antigenic structure they carry. To reach a better understanding of the nature and conformation of the ribosome-bound immunoreactive chains quantitative experiments were therefore necessary. This was carried out using a different protein/antibody system, investigated in our laboratory for several years, which seemed better adapted to such studies.

3. QUANTITATIVE ANALYSIS OF RIBOSOME-BOUND IMMUNOREACTIVE TRYPTOPHANE-SYNTHASE β-CHAINS

Extensive investigation of β-chain immunoreactivity during refolding *in vitro* showed that the monoclonal antibody mAb 19 was able to react with the refolding chains only after they had undergone some early folding steps (Murry-Brelier & Goldberg 1988; Blond-Elguindi & Goldberg 1990). Using this antibody, we investigated the immunoreactivity of ribosome-bound nascent β-chains obtained *in vitro* in a cell-free protein biosynthesis system; the general strategy was as follows (Fedorov *et al.* 1992).

Protein synthesis was achieved using either a mRNA carrying the entire *trpB* gene or a truncated mRNA that terminates, without a stop codon, at position 309 of that gene. ^{35}S-methionine was used to radiolabel the newly synthesized polypeptide chains. The ribosomal fraction was purified by centrifugation on a sucrose or glycerol gradient, and was incubated with Sepharose beads coupled to mAb19. The beads were pelleted by low speed centrifugation and washed; the immunoadsorbed material was submitted to SDS-PAGE. Autoradiography revealed the presence of radioactive material specifically immunoadsorbed by mAb 19. The shortest nascent chains that could be seen on the gel had a molecular mass of approximately 11.5 kDa. Moreover, using a newly developed radioimmunoassay (RIA) based method (Friguet *et al.* 1993), the affinity of the ribosome-bound 11.5 kDa chains for mAb 19 could be measured. It was also found to be similar to that of the native protein for mAb19. It was therefore concluded that ribosome-bound polypeptides can fold, even before the synthesis of the complete N-terminal domain of the β-chain, into a conformation that already exhibits some local structural features of the native protein and of folding intermediates observed *in vitro* (Fedorov *et al.* 1992).

4. QUANTITATIVE PULSED IMMUNOLABELLING OF GROWING POLYPEPTIDE CHAINS

In the previously described experiments, the conformation of the nascent chains was probed with the mAb well after the arrest of their synthesis. It was, therefore, impossible to ascertain whether the folded intermediate appeared rapidly while the chains were growing, or slowly after their synthesis. To overcome this difficulty, we developed a procedure to pulse-label the protein rapidly with mAb 19 during its synthesis, and quantitatively analyse the labelled, ribosome-bound, nascent chains.

While developing this procedure, several unexpected observations were made; since they had to be accounted for while setting up a correct pulsed immunolabelling protocol, we shall discuss them here. ^{35}S-Methionine labelled nascent chains were produced *in vitro* with a wheat germ translation system using a truncated mRNA that terminates, without a stop codon, at position 1104 of the *trpB* gene. This mRNA encodes the 368 N-terminal residues of the β-chain.

The absence of a stop codon at its end was supposed to ensure a stable association of these chains with the ribosomes and one therefore expected the accumulation of nascent chains of about 40 kDa which should not be released from the ribosomes. To test this prediction, the synthesis mixture was submitted to SDS-PAGE, followed by scanning the ^{35}S radioactivity in the gels with a recently developed bi-dimensional radioactivity scanner that detects the weak electrons emitted by ^{35}S or ^{14}C with very high sensitivity and very low background (β-Imager: Biospace Instruments, Thoiry, France). The results of such scans showed the accumulation of chains of the predicted length (approximately 40 kDa), but also the accumulation of shorter chains that gave rise to a distinct band pattern on the gels. Such bands are usually interpreted as reflecting transient intermediates that correspond to translation pauses at specific sites, often thought to be rare codons. The persistence of radioactive chains of intermediate lengths after a chase experiment, the demonstration that no post-translational degradation of the polypeptide chains had occurred and a quantitative analysis of the kinetics of accumulation of these bands and of the full length products allowed us to rule out this generally accepted interpretation unambiguously. It was then demonstrated that the bands of intermediate lengths in fact correspond to abortive translation products resulting from degradation of the mRNA at preferential sites.

To verify the prediction that the absence of a stop codon at the end of the mRNA should prevent the release of the nascent chains from the ribosomes, the synthesis mixture was submitted to ultracentrifugation through a glycerol cushion (Fedorov *et al.* 1992), and the resulting pellet (ribosomal fraction) and supernatant (free chains) were submitted to SDS-PAGE. Quantitative scanning of the gels with the β-Imager showed that only about half of the chains remained attached to the ribosomes, whereas the other half was released in the solution. An even lower proportion of the nascent chains (about 30%) was found bound to the ribosomal fraction when an *E. coli* translation system was used in similar experiments. This demonstrated that the commonly used method to produce ribosome-bound nascent chains, i.e. the use of a mRNA devoid of a stop codon, does not ascertain that all nascent chains remain associated to the ribosome.

To eliminate the free chains when probing the immunoreactivity of ribosome-bound nascent chains, a new separation method, rapid and compatible with the small sample volumes used (40 μl) had therefore to be devised. This was achieved by centrifugation of the sample of synthesis mixture through a 100 μl cushion of 40% glycerol in buffer A (20 mM Hepes pH 7.6, 100 mM K-acetate, 10 mM Mg-acetate) in an air-driven bench centrifuge (Airfuge, Beckman) kept at 4 °C in the cold room. The air pressure was adjusted at 30 psi, which corresponded to 149 000 g at the bottom of the centrifugation tube: 30 min of centrifugation was deemed sufficient to pellet all the ribosomal fraction and leave the free chains in solution. It was then observed that the chains thus obtained in the ribosomal fraction remain bound to the ribosomes after further incubation at 4 °C. This separation protocol, which will probably become of general use in studies on nascent proteins, was used throughout the work that is now described.

A pulse-immunolabelling protocol, compatible with the additional constraints imposed by the separation procedure just described, was set up. Aliquots of the translation mixture were supplemented with mAb 19 and incubated for 30 s. To prevent subsequent association of the antibody with nascent chains not yet folded at the time of the 30 s pulse but that would fold at a later stage, an excess of pure native non radioactive β_2 was quickly added to block the unreacted mAb. The sample was incubated with gentle shaking for 30 s, and the temperature kept at 4 °C to minimize dissociation of bound mAb 19. Immunolabelled chains were then precipitated with protein G-Sepharose beads, solubilized, submitted to SDS-PAGE and analyzed quantitatively with the β-Imager. The specificity of the immunoprecipitation had been demonstrated earlier, using a non specific mAb instead of mAb 19 (Fedorov *et al.* 1992); its linearity was verified by immunoprecipitation and quantification of the nascent chains contained in aliquots of different volumes of an arrested synthesis mixture. It was observed that, under these pulse and immunoprecipitation conditions, the fraction of chains immunoprecipitated was indeed constant ($9 \pm 2\%$, determined from four independent experiments).

This pulse-labelling and immunoprecipitation protocol was applied to aliquots of a synthesis mixture at different times after the start of synthesis. The results of such experiments are shown in figure 1*a*. Visual inspection of the screen image suggested that the appearance of immunoprecipitated (i.e. folded) chains strictly paralleled protein synthesis; a quantitative analysis of the radioactivity contained in each band was easily made using the β-Imager software. Such analysis provided the kinetics of appearance of newly synthesized (folded or not folded) chains of a given length and thus, chains of about 11, 30 and 40 kDa were shown to appear linearly with time (i.e. each at a constant rate) after a lag that increased with the chain length (see figure 1*b*). For a given chain length, this lag probably represents the time it takes for a ribosome to progress on the mRNA from the point of initiation of translation to the codon corresponding to the end of the polypeptide chain.

The kinetics of accumulation of all the ribosome-bound chains of a given size, whether immunoreactive or not, were then compared with the kinetics of accumulation of only those ribosome-bound chains of that size that were immunoreactive at the time of the pulse. These kinetics could not be distinguished from one another within the experimental conditions. The radioactivity detected in each band after pulse-labelling and immunoprecipitation represented in all cases about 10% of the total radioactivity initially contained in that band (i.e. before immunoprecipitation). This value was close to that obtained when testing the linearity and efficiency of the immunolabelling/precipitation protocol, well after the arrest of synthesis.

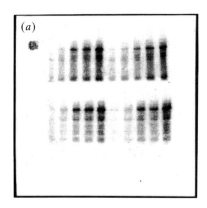

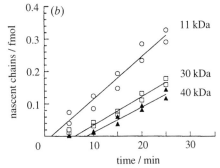

Figure 1. Kinetics of appearance of ribosome-bound ^{35}S-methionine labelled chains. Protein synthesis was performed in a wheat germ translation system, using as a template a mRNA that ends, without a stop codon, at base 1104 of the *trpB* gene. Aliquots were withdrawn after various times and synthesis was arrested by dilution with cold buffer. The ribosomal fraction was separated by Airfuge centrifugation through a glycerol gradient (see text). The pellet was resuspended in sample SDS buffer, heated, and each sample was subjected to SDS-PAGE. (*a*) Screen image of the radioactivity distribution in the gel for the chains present before (top) or after (bottom) immunoprecipitation. Samples used to measure the total synthesis before immunoprecipitation (15 µl, top half) correspond to 1/10 of those used for the immunoprecipitation (bottom half). Assays were performed in duplicate, after 5 min (lanes 1, 6), 10 min (lanes 2, 7), 15 min (lanes 3, 8), 20 min (lanes 4, 9), and 25 min (lanes 5, 10) of incubation of the translation mixture at 25 °C, and were repeated in three independent experiments. A ^{14}C sample of known radioactivity (500 d.p.m.) served as an internal standard (top left hand corner). (*b*) In each migration lane (i.e. for each incubation time) the radioactivity contained in the region corresponding to polypeptide chains of about 11, 30 and 40 kDa was determined by means of the β-Imager software and was converted into fmoles of the corresponding polypeptide chains, taking into account the internal standard, the specific radioactivity of the ^{35}S-methionine used, and the number of Met residues present in each fragment. Based on the sequence of the *E. coli trpB* gene, there are 15, 10 and 4 Met residues in the 40, 30 and 11 kDa fragments respectively. The lags observed for the biosynthesis of the β-chain fragments were 1.5 min (11 kDa fragment, open circles), 6 min (30 kDa fragment, open squares), and 8.5 min (40 kDa fragment, filled triangles). Data were fitted by least squares linear regression, taking into consideration measurements done after 10 min for the 30 and 40 kDa fragments, and including those for 5 min for the 11 kDa fragment.

This result suggested that the antigenic site recognized by mAb 19 was completely immunoreactive at the moment of the 30 s immunopulse; moreover, the lags observed for the accumulation of total and immunoreactive 40 kDa ribosome-bound chains were very similar, the difference not exceeding the 30 s duration of the pulse. This observation strongly supported the conclusion that the appearance of the antigenic site on the 40 kDa nascent chains was completed at the time of the pulse. The same conclusion could be reached with the 11 and 30 kDa fragments which also showed a linear accumulation of immunoreactive ribosome-bound chains with a lag not significantly different from that observed for the synthesis. Because, under the experimental conditions we used, it took the ribosome about five minutes to elongate a nascent chain from 11–40 kDa (as judged from the lags observed in figure 1*b*), these results unambiguously showed that the antigenic site recognized by mAb 19 appears cotranslationally.

Because studies on the refolding *in vitro* of unfolded β-chains had demonstrated that the appearance of immunoreactivity requires some early folding steps, one was tempted to conclude that these pulsed immunolabelling experiments solidly demonstrated that nascent β-chains undergo these folding steps cotranslationally. A detailed characterization of the antigenic site showed that such a conclusion would have been premature.

5. CHARACTERIZATION OF THE EPITOPE RECOGNIZED BY ANTIBODY 19

To understand the nature of the early folding steps detected *in vitro* by mAb19, it seemed desirable to define as precisely as possible the residues involved in the corresponding antigenic determinant. This was done by a variety of techniques including DNA sequencing of an epitope library, enzymatic and chemical cleavage of β_2 at specific sites, and chemical peptide synthesis. The epitope recognized by mAb19 has thus been localized in the aminoacid sequence 2–9 of the β-chain. The affinities of mAb 19 for several synthetic peptides of different lengths containing the 2–9 sequence have been determined and were all close to 10^9 M^{-1}, the affinity of mAb19 for the native β_2 protein. Because an isolated peptide of ten residues is very unlikely to adopt a prefered rigid conformation, this suggested mAb 19 may recognize the octapeptide 2–9 even in a flexible disordered state. To account for the fact that mAb19 fails to react with β-chains before they have undergone some folding steps, we propose that, at the very beginning of the *in vitro* folding process, the antigenic N-terminal end of the polypeptide chain (which turns out to be very hydrophobic) would get very rapidly, but transiently buried in the hydrophobic interior of a molten globule. At a later stage of the folding process, upon compaction of the molten globule, the 2–9 sequence would be expelled towards the solvent and only then would its immunoreactivity to mAb19 become apparent. It should be noted that, *in vitro*, the antigenic site recognized by mAb 19 was shown to appear after the formation of a molten globule, but before the final compaction of the protein core (Goldberg *et al.* 1990).

Thus, mAb 19 appears capable of recognizing either unfolded β-chains, or chains that have already gone some way along the folding pathway, but not molten globule-like intermediates situated early on the folding pathway.

6. ASSOCIATION OF NEWLY SYNTHESIZED β CHAIN FRAGMENTS WITH CHAPERONES

Because mAb19 can recognize unfolded β-chains, the identification by pulsed immunolabelling of immunoreactive nascent chains with this mAb could no longer be considered as evidence that these chains had undergone some folding steps. Rather, it could be envisaged that, by preventing the nascent chains from undergoing a hydrophobic collapse before their release from the ribosomes, chaperones might maintain the nascent chains in an immunoreactive state. We therefore investigated the possibility that chaperones might bind efficiently to nascent β-chains during *in vitro* biosynthesis.

First, a quantitative detection of GroEL and of DnaK on Western blots was set up using commercially available monoclonal antibodies specific of these two proteins; anti-mouse immunoglobulin antibodies labelled with ^{35}S and the β-imager radioactivity scanner. This allowed us to estimate the concentrations of endogenous GroEL and DnaK as being about 0.25 and 1 mg/ml, respectively in the *E. coli* extract used for cell-free protein synthesis. These concentrations of chaperones are certainly large enough to allow for significant association with proteins in the process of folding.

To detect a possible association of chaperones with newly synthesized polypeptide chains, either bound to, or released from the ribosomes, the following experiments were conducted. Protein synthesis was allowed to proceed *in vitro* for 60 min at 37 °C, using an *E. coli* extract, ^{35}S-methionine, and the truncated mRNA that terminates at position 1104 without a stop codon. The newly synthesized chains, released and ribosome-bound were separated from each other by centrifugation on a glycerol cushion as described above. Chains released in the supernatant were immediately subjected to electrophoresis on a non-denaturing, 6–10% acrylamide-gradient gel according to Lambin & Fine (1979), under conditions where the migration of the proteins depend essentially on their molecular mass. After migration, staining and drying, the radioactivity distribution in the gel was analysed by use of the β-imager; a typical recording is shown in figure 2. The diagram obtained shows that the newly synthesized polypeptide chains are all found in high molecular mass complexes, distributed in two peaks. The first, very sharp peak migrates with an apparent molecular mass of approximately 900 kDa, at a position which exactly coincides with that of GroEL: it was detected using either an anti-GroEL monoclonal antibody to Western blot a lane of the gel that contained the whole translation mixture, or by

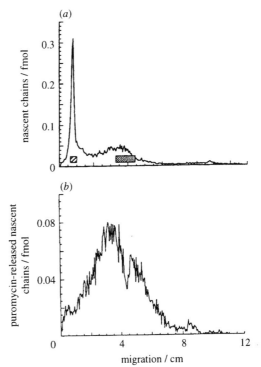

Figure 2. Association of ^{35}S-labelled newly synthesized chains with chaperones revealed by electrophoresis in acrylamide gradient non denaturing gels. ^{35}S-methionine labelled chains were synthesized in an *E. coli* translation system using the same mRNA as in figure 1. (*a*) The ribosomal fraction was removed by Airfuge centrifugation through a glycerol cushion. The ribosome-free fraction was submitted to electrophoresis on a 6–10% acrylamide gradient nondenaturing gel. Radioactivity was recorded in the β-imager, and was plotted (in amount of synthesized nascent chains per 10 μl of translation mixture) as a function of the migration in the gel. The bars indicate the position of a band revealed by anti GroEL (hatched shading) or antiDnaK (stippled shading) antibodies in Western blotting experiments. (*b*) The ribosome-bound nascent chains were separated by centrifugation through a glycerol cushion in the Airfuge and resuspended in 50 μl of buffer (20 mM Hepes pH 7.6, 100 mM K-acetate and 10 mM Mg-acetate) containing 10 mM puromycin and 1 M KCl. After 30 min incubation at 25 °C, the sample was freed of ribosomes by a second Airfuge centrifugation. An aliquot of the supernatant, corresponding to 25 μl of the initial translation mixture and containing the radioactive nascent chains that were released from the ribosomes by the puromycin treatment was submitted to electrophoresis on a nondenaturing 6–10% acrylamide gradient gel. After staining and drying, the gel was analysed in the β-imager and the radioactivity was plotted as a function of the migration in the gel.

Coomassie blue staining of a lane of the gel that contained only pure GroEL. The second, much wider, peak which corresponded to apparently heterogeneous species, migrated in a region of the gel where DnaK could be detected by Coomassie blue staining in a lane with the pure protein, and by Western blot with an anti-DnaK monoclonal antibody in a lane containing the nascent chain preparation. Thus, newly synthesized incomplete polypeptide chains released from the ribosomes were found in stable high molecular mass complexes containing either GroEL or DnaK, perhaps associated with other chaperones.

To discover whether these complexes with chaperones were formed before or after the release of the nascent chains from the ribosomes, the same approach (i.e. electrophoresis in a non-denaturing, acrylamide gradient gel) was applied to the ribosomal fraction obtained after the Airfuge centrifugation (see previous paragraph). However, before electrophoresis, the nascent chains had to be released from the ribosomes; to permit their migration in the gel; this was achieved by incubation of the resuspended pellet with 10 mM puromycin and 1 M KCl for 1 h at 37 °C (conditions mild enough to respect the integrity of the ribosomes; Blobel & Sabatini 1971). Under these conditions, about 25 % of the ribosome-bound nascent chains were released in solution and were found in the soluble fraction after a second Airfuge centrifugation.

This fraction was subjected to non-denaturing electrophoresis; the gel was stained, dried and analysed in the β-imager. Unlike what had been observed for chains spontaneously released from the ribosomes, the radioactivity pattern obtained with the puromycin released chains did not show the sharp peak corresponding to a complex with GroEL. On the contrary, the broad peak corresponding to a complex containing DnaK was also present with the puromycin released chains; in addition, no radioactivity was detected in the region of the gel corresponding to free (i.e. not bound to DnaK) polypeptide chains. Because no free chaperones were present in the solution at the time of, or after, the puromycin treatment, these observations demonstrated that DnaK was stably bound to the nascent chains before their release from the ribosomes, and that no nascent chain of significant length was found free of chaperone. But although GroEL was not found in a ribosome-bound complex, it does not exclude the possibility that such complexes exist: dissociation could have occured during the puromycin treatment and centrifugation step. Thus, our results demonstrate that if they exist, complexes of GroEL with ribosome-bound nascent chains (unlike the complexes with chains released from the ribosomes) should be rather unstable.

7. CONCLUSION

The results described in this communication deal with two distinct, though not unrelated, problems: protein synthesis *in vitro* and protein folding during elongation; they can be summarized as follows. Concerning protein synthesis, we showed the following.

1. In the systems we used at least, the pattern of discrete bands observed on SDS gels does not reflect transient translation pauses. Rather, it results from the presence of premature translation termination caused by the degradation of the mRNA at preferential sites.

2. The absence of a stop codon at the end of the mRNA does not ensure that the nascent chains will remain ribosome bound. Indeed, a large fraction of the chains were released in solution in both a wheat germ and an *E. coli* translation system. This lead us to develop a small-scale method for separation of the ribosome-bound from the free chains.

3. The efficiency of the translation system was quantitatively determined by measuring the number of full length chains produced.

4. The concentrations of endogenous GroEL and DnaK in cell-free translation systems were determined.

Concerning the folding problem, we showed the following.

1. For the phage P22 tailspike protein a specific monoclonal antibody detecting structured monomers during the folding *in vitro* can bind to ribosome-bound nascent chains. This suggests that folding steps that take place in solution during *in vitro* refolding also occur in a ribosome-bound state *in vivo*.

2. In the case of the β-chain of *E. coli* tryptophan-synthase, the use of a short immunochemical pulse-labelling method demonstrated that the appearance of the immunoreactivity is cotranslational.

3. Newly synthesized β-chains or β-chain fragments were always found in high molecular complexes. The chains released from the ribosomes were found in complexes containing either GroEL or DnaK. The ribosome-bound nascent chains were found in complexes with DnaK, but not with GroEL.

4. The epitope recognized by the antibody (mAb19) used for the immunochemical labelling was characterized. Its properties suggest that only some collapsed early folding intermediates fail to interact with mAb19. Conversely, unfolded β-chains or β-chain fragments, as well as chains that have already progressed some way along the folding pathway, should bind to the antibody. Thus the immunoreactivity to mAb19 is not a direct proof that folding has occurred.

That all ribosome-bound nascent β-chains were found associated to DnaK suggests the possibility that this chaperone (and perhaps others) may prevent the polypeptide chains from undergoing the initial folding steps that lead to transiently bury residues 2–9, and hence maintain them in a state where they are recognized by mAb19. Therefore, in view of the recently discovered properties of mAb19, our previous conclusion (Fedorov *et al.* 1992) that nascent chains can start their folding on the ribosome was premature, and the question of whether or not folding starts cotranslationally remains unanswered. Nevertheless, using the experimental approach described here together with a mAb that would recognize a discontinuous epitope should bring an answer to this question.

REFERENCES

Blobel, G. & Sabatini, D. 1971 Dissociation of mammalian polyribosomes into subunits by puromycin. *Proc. natn. Acad. Sci. U.S.A.* **68**, 390–394.

Blond-Elguindi, S. & Goldberg, M.E. 1990 Kinetic characterization of early immunoreactive folding intermediates during the renaturation of guanidine-unfolded *Escherichia coli* tryptophan synthase β_2 subunits. *Biochemistry* **29**, 2409–2417.

Fedorov, A.N., Friguet, B., Djavadi-Ohaniance, L., Alakhov, Y.B. & Goldberg, M.E. 1992 Folding on the ribosome of *Escherichia coli* tryptophan synthase β subunit nascent chains probed with a conformation-dependent monoclonal antibody. *J. molec. Biol.* **228**, 351–358.

Friguet, B., Djavadi-Ohaniance, L., Haase-Pettingel, C.A.,

King, J. & Goldberg, M.E. 1990 Properties of Mono-clonal Antibodies Selected for Probing the Conformation of Wild Type and Mutant Forms of the P22 Tailspike Endorhamnosidase. *J. biol. Chem.* **265**, 10347–10351.

Friguet, B., Fedorov, A., Serganov, A.A., Navon, A. & Goldberg, M.E. 1993 A radio-immuno-assay based method for measuring the true affinity of a monoclonal antibody with trace amounts of radioactive antigen. Illustration with the products of a cell-free protein synthesis system. *Analyt. Biochem.* **210**, 344–350.

Friguet, B., Djavadi-Ohaniance, L., King, J. & Goldberg, M.E. 1994 In vitro and Ribosome-bound Folding Intermediates of P22 Tailspike Protein Detected with Monoclonal Antibodies *J. biol. Chem.* **269**, 15945–15949.

Goldberg, M.E., Semisotnov, G.V., Friguet, B., Kuwajima, K., Ptitsyn, O.B. & Sugai, S. 1990 An early immuno-reactive folding intermediate of the tryptophan synthase β_2 subunit is a 'molten globule'. *FEBS Lett.* **263**, 51–56.

Goldenberg, D.P. & King, J. 1982 Trimeric intermediate in the in vivo folding and subunit assembly of the tailspike endorhamnosidase of bacteriophage P22. *Proc. natn. Acad. Sci. U.S.A.* **79**, 3403–3407.

Haase-Pettingel, C.A. & King, J. 1988 Formation of aggregates from a thermolabile in vivo folding inter-mediate in P22 tailspike renaturation. *J. biol. Chem.* **263**, 4977–4983.

Lambin, P. & Fine, J.M. 1979 Molecular weight estimation of proteins by electrophoresis in linear polyacrylamide gradient gels in the absence of denaturing agents. *Analyt. Biochem.* **98**, 160–168.

Murry-Brelier, A. & Goldberg, M.E. 1988 Kinetics of appearance of an early immunoreactive species during the refolding of acid-denatured *Escherichia coli* tryptophan synthase β_2 subunit. *Biochemistry* **27**, 7633–7640.

Seckler, R., Fuchs, A., King, J. & Jaenicke, R. 1989 Reconstitution of the thermostable trimeric phage P22 tailspike protein from denatured chains in vitro. *J. biol. Chem.* **264**, 11750–11753.

Folding and association versus misfolding and aggregation of proteins

RAINER JAENICKE

Institut für Biophysik und Physikalische Biochemie, Universität Regensburg, D-93040 Regensburg, Germany

SUMMARY

The acquisition of spatial structure in proteins may be described in terms of hierarchical condensation, with contributions of local interactions between next neighbours and the interactions between domains and subunits accumulating to create the marginal free energy of stabilization characteristic of the functional state of globular proteins. Domains represent independent folding units such that the overall kinetics divide into the sequential collapse of subdomains and domains and their merging to form the compact tertiary structure. In proceeding to oligomeric proteins, docking of subunits follows the formation of structured monomers. Thus, the overall mechanism of folding and association obeys consecutive uni–bimolecular kinetics. Beyond a limiting protein concentration, aggregation will outrun proper domain pairing and subunit association. In the cell, accessory proteins are involved in catalysis of the rate-determining steps of folding (proline isomerization and SH–SS exchange) and in the kinetic partitioning between folding and aggregation (chaperone action). The practical aspects of accessory proteins have been investigated in detail using immunotoxins and antibody fragments as examples. Additional concepts allowing off-pathway reactions in protein reconstruction to be kept to a minimum refer to pulse-dilation, reverse micelles and immobilization of polypeptide chains on matrices.

1. INTRODUCTION

The topic of this paper has three 'faces': one looking back into the heroic age when Anson for the first time observed 'reversible denaturation' of proteins; a second one facing industrial applications; and the third one with a vision of how kinetic partitioning between folding and misfolding and their consecutive reactions in the cell might work.

With respect to the first issue, the correlation between hierarchies of protein structure, stability and folding, and recent experimental developments allowing access to fast and/or local events in the course of protein self-organization, must be considered. Regarding the second issue, what started 15 years ago in basic research has developed to a huge extent: manual mixing of microlitre aliquots has changed into cubic metres of 6 M guanidine solutions, pumped in automated processes at industrial plants to yield several grammes of pharmaceutical product at a time. To the *in vitro* versus *in vivo* controversy the physical biochemist cannot contribute much, except the reductionist's scepticism against cartoons which make readers (and sometimes authors) believe that the cell is full of circles and triangles instead of molecules. What is required, and becomes increasingly accessible due to molecular biological techniques, is the structural and functional analysis of components involved in chaperoning and targeting the nascent polypeptide chain. In addition, experimental approaches have to be devised which allow cellular conditions to be mimicked. Significant progress has been made toward the goal of reproducing *in vivo* protein folding in the test tube. Examples include studies of folding in the presence and absence of accessory proteins, complementation experiments using transcription–translation systems and *in vitro* translation of mRNAs encoding truncated or otherwise mutated mRNAs.

2. HIERARCHIES OF FOLDING AND STABILITY

In the structural hierarchy of proteins, the different levels refer to folding as well as stability. Increasing packing density and release of water from hydrophobic residues provide the enthalpic and entropic increments of the free energy of stabilization which finally yield the marginal difference of the attractive and repulsive forces characteristic for the native-state stability of biological macromolecules. Typical ΔG_{stab} values are approximately 50 kJ mol^{-1}; this is equivalent to just a few weak interactions in a protein molecule with *ca.* 5000 atoms. The biological significance of this observation is three-fold: (i) optimization of the structure–function relation in the course of evolution is aimed at flexibility (catalysis, regulation, turnover) rather than stability; (ii) under physiological conditions, native globular proteins are generally on the borderline of denaturation; and (iii) because the native state is a state of minimum potential energy, folding intermediates must exhibit even lower stability than their native counterparts, so misfolding and subsequent kinetic competition of reshuffling and off-pathway reactions are expected to occur (Jaenicke 1991a, b, 1993a; Jaenicke & Buchner 1993). Evidence that protein biosynthesis and folding in the cell are not 100 % efficient came from turnover experiments. For

Phil. Trans. R. Soc. Lond. B (1995) **348**, 97–105
Printed in Great Britain

97

© 1995 The Royal Society

example, even under optimum growth conditions within the host, the tail spike protein of phage P22 has a yield of less than 50%. Under unbalanced physiological conditions, only proteins with incorrect conformations are produced, then continuously removed by proteolysis.

In correlating the intrinsic stability of proteins with the above hierarchy, thermodynamic measurements on point mutants, protein fragments and homologues differing in their state of association have clearly shown that each structural level makes its own contribution. As shown by NMR and other spectroscopic techniques, oligopeptides may form stable non-random conformations; at a minimum length of 15 residues, they have been shown to sustain native-like structure (Baldwin 1991). Their thermal unfolding–refolding behaviour can be quantitatively described by the standard helix–coil theories, even for short peptides. This is because both the helix nucleation constant and the enthalpy change per mole residue for helix formation are insensitive to the length of the polypeptide chain. Regarding larger fragments, it has long been known that both subdomains and domains exhibit high intrinsic stabilities which are not too different from the free energies observed for the uncleaved parent molecule. In numerous cases significant mutual stabilization has been observed both at the subdomain level and at the level of domains and subunits. A striking example is the mutual stabilization of the N- and C-terminal domains of γB-crystallin (Mayr *et al.* 1994), where the complete molecule shows the typical bimodal equilibrium transition, with the second phase superimposable over the unfolding transition of the isolated N-terminal domain fragment, whereas the isolated C-terminal domain shows surprisingly low intrinsic stability.

The mutual stabilization of subunits is illustrated by lactate dehydrogenase where the stability decreases steadily from the highly stable (native) tetramer down to domain fragments: the 'proteolytic dimer' requires structure-making salts to exhibit activity, whereas the 'structured monomer' is inactive and can only be detected as a short-lived folding intermediate on the pathway of reconstitution. The separate NAD- and substrate-binding domains are unstable but still sufficiently structured to recognize each other, exhibiting a mutual 'chaperone effect' during joint renaturation (Opitz *et al.* 1987). Extrinsic factors such as ions, cofactors and non-proteinaceous components (e.g. nucleic acids or carbohydrates) may contribute significantly to protein stability; they are also important in determining the mechanism of folding and the state of association (Jaenicke 1987).

3. MECHANISM OF FOLDING AND ASSOCIATION

The fact that protein folding must occur within biologically feasible time (i.e. a timespan which is much shorter than an organism's lifetime) excludes the possible existence of a random-search mechanism over all conformational space, forcing the assumption that there must be kinetic pathways for folding.

Advances in spectroscopic methodology laid a foundation for the understanding of ordered pathways and well defined intermediates. Starting from local, next-neighbour interactions, 'seeds' in the folding proceed to rate-determining late events such as proline isomerization and disulphide formation. Small proteins or constituent domains of large proteins commonly collapse, in a highly cooperative manner, into a compact structure. This structure is usually native-like, although there have been reports showing that non-native intermediates may occur. As a general principle, fast secondary structure formation precedes slow multistep rearrangements, observable on the seconds to minutes timescale. However, due to limited time resolution and problems associated with multi-step and multiple pathways, it has not yet been possible to elucidate the complete timecourse of folding. Currently, the most detailed mechanisms available refer to small single-chain one-domain proteins: basic pancreatic trypsin inhibitor (BPTI), ribonuclease, hen egg-white lysozyme, and barnase (cf. Jaenicke & Buchner 1993).

The following general conclusions have been derived from these model systems: (i) in agreement with the above two-phase mechanism, there are compulsory pathways of folding which are, at least in part, sequential; (ii) secondary structure formation is driven by local hydrophobic surface minimization and precedes tertiary structure formation; and (iii) tertiary interactions become increasingly defined as water release consolidates the hydrophobic core.

In proceeding to domain proteins and protein assemblies, the previous conclusions remain widely unchanged. This is because proteins 'fold by parts' i.e. the domains fold and unfold independently according to

$$N_i - N_j \rightleftharpoons N_i - U_j \rightleftharpoons U_i - U_j, \qquad (1)$$

where i and j refer to different domains in their native (N) and unfolded (U) states. (This observation is true both *in vitro* and in the cell; Jaenicke 1993*a*.) The mechanism is significant with respect not only to protection of the nascent polypeptide chain from proteases, and the evolution of multifunctional enzymes, but also as a contribution to the rate enhancement of protein folding.

There are examples where biological function requires the cooperation of domains; for example, two domains forming one active centre. In such cases, domain pairing may occur as an additional step in the overall folding reaction, whereas domain folding represents a precursor reaction. In such cases, the above folding reaction (see equation 1) may contain domain pairing as an additional step. For a detailed analysis of sequential folding over the time range from milliseconds to seconds, see Blond-Elguindi & Goldberg (1990).

In oligomeric proteins, subunit assembly corresponds to domain pairing. The preceding steps on the pathway are again: formation of elements of supersecondary structure, collapse to subdomains and domains with the final formation of structured monomers, which then associate to yield the correct stoichiometry of the native quaternary structure. It is evident that the collision complex may undergo intramolecular rearrangement to reach the state of maximum packing density and minimum hydrophobic surface area. Thus, folding and association may be followed by further first-order steps so that, in the simplest case, the overall reaction of a dimer may be written as a sequential uni–bi–unimolecular reaction

$$2\mathcal{M} \longrightarrow 2M' \xrightarrow{k_1} 2M \xrightarrow{k_2} M_2 \longrightarrow N, \qquad (2)$$

with $\mathcal{M}, M', M$ as unfolded, intermediate and structured monomer, N as native dimer and k_1, k_2 as first- and second-order rate constants (Jaenicke 1987).

Monitoring of single steps along the pathway of folding and association is dependent on the specific structure–function relation for a given quaternary structure. In most cases, biological function requires the native state of association so that the final rate-determining step can be monitored by measuring activity. Preceding steps may be accessible to spectral analysis, crosslinking, HPLC and a wealth of other methods (Jaenicke & Rudolph 1988).

The classical repertoire of methods does not allow analysis of the initial fast phase of protein folding although, recently, folding reactions in the sub-milliseconds time range became accessible, using time-resolved absorption spectroscopy (Jones *et al.* 1993).

4. OFF-PATHWAY REACTIONS

There are three stages where side reactions on the folding pathway may compete with proper folding and association: hydrophobic collapse; merging and 'swapping' of domains; and docking of subunits. In all cases recognition is involved, in the sense that specific substructures or surfaces must be preformed so that folding can proceed towards maximum packing density and minimum hydrophobic surface area.

Both hydrophobic collapse and domain merging involve intramolecular rearrangements. Owing to the high local concentration of the reacting groups or surfaces within one and the same polypeptide chain, they are not significantly affected by neighbouring molecules: i.e. they obey first-order kinetics, the slowest isomerization reaction determines the overall rate. In the case of domain proteins, the relative stabilities of the domains and the contributions of the domain interactions to the overall stability are crucial. The significance of the linker peptide connecting two well-defined domains has been studied by using grafting experiments; for example, mutual exchange of the linkers of β- and γ-crystallin (Jaenicke 1994). Domain contacts dominate over subunit contacts in both transplants. The recombinant (separate) domains do not interact with each other, stressing the above-mentioned argument for local concentration. The suggestion that separate domains or other substructures may recognize each other and form complexes has been known ever since the discovery of RNaseS. True domain recognition, and the folding and association of nicked polypeptide chains in lactate dehydrogenase, have been studied in detail by Opitz *et al.* (1987). The fact that, in this case, the yield does not exceed 15% illustrates the importance of side reactions (see below).

Kinetic partitioning upon reactivation of the related monomeric octopine dehydrogenase (ODH), yields 70% intact ODH and 30% inactive protein with native-like secondary structure, but increased hydrophobicity (Teschner *et al.* 1987):

$$\mathcal{M} \longrightarrow U \longrightarrow I_N \longrightarrow N \quad (70\%)$$
$$\downarrow$$
$$I_N^* \qquad (30\%). \qquad (3)$$

N represents the native state characterized by correct domain pairing, whereas I_N^* is trapped, probably as a consequence of wrong domain interactions and subsequent aggregation.

In going from of single-chain domain proteins to protein assemblies (with n subunits), kinetic competition of folding and association comes into play. This is because subunit association requires that the monomers are close to their proper conformation before they coalesce to form the native quaternary structure. If folding intermediates expose wrong hydrophobic contact sites, aggregation rather than association will result, and because now bimolecular steps are involved in the process, the second-order kinetics of association are enhanced at high protein concentration. The occurrence of inclusion bodies as a common feature of recombinant protein technology illustrates the consequences. The underlying kinetic mechanism

$$n\mathcal{M} \longrightarrow nM' \xrightarrow{k_1} nM \xrightarrow{k_2} M_n$$
$$\downarrow \qquad \downarrow$$
$$\mathcal{M}_n \qquad M_n' \qquad\qquad (4)$$

(with $\mathcal{M}, M', M$ as unfolded, collapsed and structured monomers, and k_1, k_2 as first- and second-order rate constants) corresponds to equation (3) which represents the limiting case at high dilution; it is sufficient to quantitate the observed inverse concentration dependence of reactivation and aggregation (Kiefhaber *et al.* 1991).

Taking equation (4) into consideration, three questions need to be answered: (i) what is the committed step in aggregate formation; (ii) when is the structured monomer committed to end up as the native protein; and (iii) what is known about the structure of aggregates and their constituent polypeptide chains? Regarding the first two points, it was shown that commitment to aggregation was a fast reaction, whereas the kinetics of the 'commitment to renaturation' followed precisely the slow kinetics of overall reactivation. This means that there are fast precursor reactions on the folding path (collapsed states) which still permit aggregation, whereas, after a certain intermediate has been formed, slow shuffling leads 'one way' to the native state (Goldberg *et al.* 1991). Concerning the structure of aggregates, electron microscopy and circular dichroism show that wrong subunit interactions give rise to irregular networks with a broad distribution of highly structured particles at least ten times the size of the native proteins. They resemble the native protein in its spectral properties,

Table 1. *Characterization of heart-muscle lactate dehydrogenase in its native, denatured and aggregated states*

	native state	denatured state	aggregated state
	pH 7.6	pH 2.3	pH 7.6
specific activity/(U mg^{-1}) ,	639 ± 40	0	0
maximum fluorescence/nm	299 ± 1	333	331
$-[\theta]$222 nm (deg cm^2 dmol^{-1})^a	15300 ± 1000	10800	14700
sedimentation coefficient/S	7.6 ± 0.2	2.7	$\geqslant 20$
molecular mass/Da^b	140000 ± 1500	35000	~ 1000000
$k_{2,\text{react}} \times 10^{-3}/(\text{M}^{-1} \text{c}^{-1})^c$			
10 mM GdmCl	—	11.4	11.6
30 mM GdmCl	—	4.3	4.7
120 mM GdmCl	—	0.92	0.81

^{a} Ellipticity at 222 nm.
^{b} From sedimentation equilibrium.
^{c} Second-order rate constant of reactivation at 20 °C after acid denaturation, resolubilization in 6 M guanidinium-chloride (GdmCl) at pH 2 and subsequent dilution in 0.1 M phosphate buffer pH 7.6 (residual GdmCl concentration).

as far as turbidity allows this conclusion (see table 1). For details of the structural characteristics of inclusion bodies, refer to Bowden *et al.* (1991).

5. KINETIC PARTITIONING: BIOTECHNOLOGICAL ASPECTS

(a) Aggregation and inclusion body formation

As discussed earlier, aggregation *in vitro* and formation of inclusion bodies within the cell correspond to each other; overexpression leads to high local concentrations of folding intermediates and yields precipitates instead of native protein. There are a variety of strategies to circumvent this problem.

First, one might use weaker promoters, and thus reduce the concentration. Because the aggregation reaction is above second order, this will drastically reduce the local concentration of folding intermediates, thus favouring correct folding and association. However, there are two reasons why this approach is of little practical use: first, the overall yield of the recombinant protein per gram cell mass is decreased; and second, its purification requires a full-scale separation of the guest molecule from the bulk of the host proteins, whereas purification of inclusion bodies (with their characteristic low heterogeneity) is highly simplified. Thus it is often the case that experimentation starts with the inclusion bodies; optimization focuses on *in vitro* reconstitution of the mixture after solubilization (and denaturation) in, for example, guanidinium chloride or urea. A discontinuous pulse-dilution technique has been devised to perform the dilution–reconcentration cycle in an economic way (Rudolph 1990). A specific quantity of protein is diluted and then reactivated at a concentration less than 1 µM; upon approaching the final value of reconstitution, a new aliquot of the concentrated denatured protein solution is added, and so on until the whole batch is transferred. The method has two advantages: (i) the actual concentration of the folding intermediate never exceeds the critical concentration of aggregation; and (ii) the increasing

concentration of renatured protein exerts a stabilizing effect on the folding intermediate, comparable to serum albumin which is commonly used. Additives such as arginine may strongly increase the yield by shuffling aggregates back on the productive folding path.

Little is known about specific groups involved in the aggregation reaction, but early systematic experiments suggested that, in addition to covalent disulphide linkages (Jaenicke 1967), hydrophobic interactions were of major importance. Recent *in vitro* and *in vivo* studies confirm this result (Hurtley & Helenius 1989; Mitraki & King 1989; Rudolph 1990; Helenius *et al.* 1992). The monomeric two-domain enzyme rhodanese became a test case; it is inaccessible to reactivation because of its high tendency to form aggregates, but competition by detergents for the hydrophobic aggregation sites resulted in successful renaturation (Tandon & Horowitz 1986). More detailed insights came with the study of mutants; in the case of bovine growth hormone, an extension of the hydrophobic surface was shown to result in enhanced aggregation (Brems *et al.* 1988). The partitioning between folding and aggregation has been most intensively studied in the case of the tailspike endorhamnosidase from *Salmonella* phage P22 and the numerous mutants of this protein, which have been shown to either increase or suppress aggregation (Mitraki & King 1992).

The wild-type trimer is highly stable; the *in vivo* and *in vitro* folding behaviour is very similar (Fuchs *et al.* 1991). On release from the ribosome or upon dilution from denaturant solutions, the polypeptides fold into a conformation sufficiently structured for proper assembly; most of the β-sheet secondary structure along with their aromatic amino acid side-chains are close to the native state, yet even as protrimers they are still highly unstable. The anomalous stability is only acquired in a slow rearrangement reaction when the intertwined parallel β-helices merge to form the native trimer and, correspondingly, a significant part of the tailspike folding reaction occurs after subunit association (R. Seckler, personal communication).

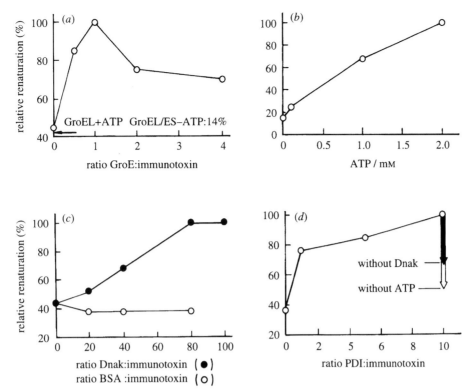

Figure 1. Renaturation of a single-chain immunotoxin facilitated by chaperones and protein disulphide isomerase. (*a*) GroE-facilitated renaturation of the GdmCl-denatured reduced protein. Optimum renaturation at equimolar GroE:immunotoxin ratio. GroEL alone has no effect; GroEL–GroES in the absence of ATP traps folding intermediates. (*b*) Effect of Mg-ATP on the yield of renaturation. (*c*) DnaK-facilitated renaturation after GdmCl-denaturation. BSA (or denatured DnaK), heated to 100 °C have no effect. (*d*) PDI-mediated oxidation of denatured and reduced immunotoxin in the presence of 60-fold excess DnaK. As indicated by the arrows, PDI and DnaK show synergistic effects (Buchner *et al.* 1992).

Both during structure formation *in vivo* and during refolding *in vitro*, the fraction of chains capable of maturing to the native form decreases with increasing temperature, with the remaining polypeptides accumulating as aggregates (Mitraki *et al.* 1993). These are formed from partly-folded intermediates that can either form native tailspike protein or form aggregates. Temperature-sensitive folding (*tsf*) point mutations reduce the folding yield at elevated temperatures, whereas second-site suppressor mutations (*su*) improve folding under such conditions (Mitraki & King 1992). Both types of mutations act by altering the stability of folding intermediates, *tsf*-substitutions destabilizing and *su*-substitutions stabilizing. In the native structure, denaturation is kinetically controlled and the effects of mutations are masked by the complicated unfolding pathway (Danner & Seckler 1993).

(*b*) Reconstitution in the presence of accessory proteins

Inclusion bodies are the product of intracellular aggregation. They differ from aggregates formed in the test tube by their high packing density and large size, which may sometimes span the entire diameter of the cell (Bowden *et al.* 1991). As a consequence they can be harvested easily and washed by fractionated centrifugation, favouring their use in the downstream processing of recombinant proteins. The protein of interest can then be isolated using the same method of

in vitro denaturation–renaturation as described before; detailed guidelines are described by Rudolph (1990).

More recently, the repertoire of methods has been extended by attempts to mimic conditions *in vivo*, with respect to folding catalysts and chaperone proteins. Two examples of this are as follows. The first deals with the reactivation of a denatured and reduced immunotoxin (B3(F$_o$)-PE38KDEL) composed of the V$_H$ region of a carcinoma-specific antibody, which is connected by a flexible linker to the corresponding V$_L$ chain, which is in turn fused to truncated Pseudomonas exotoxin. The chimeric protein contains three disulphide bonds, one in each antibody domain, and one in the toxin part. Upon renaturation, aggregation of non-native polypeptide chains and the formation of incorrect disulphide linkages create more than 90% inactive molecules. Attempts to improve the yield by adding the *Escherichia coli* chaperones GroE and DnaK and bovine protein disulphide isomerase (PDI) were successful. As illustrated in figure 1, it was discovered that both GroE and DnaK influence the reaction: GroEL alone inhibits reactivation, whereas the complete GroE-system significantly increases the yield of active protein. DnaK exhibits the same effect both in free and immobilized form, which allows the chaperone to be reused in the downstream processing of the protein. PDI stimulates reactivation synergistically, and under optimum conditions reactivation yields are doubled compared with non-enzymic disulphide bond formation (Buchner *et al.* 1992).

Table 2. *Rate acceleration of the folding of oxidized Fab fragments by PPIs*

(Apparent rate constants at 10 °C determined by fitting the kinetic data obtained by ELISA to a first-order reaction. Long-term and short-term denaturation were carried out by incubating Fab in GdmCl for > 2 h and 20 s, respectively. CP = cyclophilin; FKBP = FK506 binding protein; CsA = cyclosporin A. Data taken from Lilie *et al.* 1991.)

	$k_{app}/(min^{-1})$		
	without CP	with CP	with FKBP
long-term denaturation	0.032 ± 0.003		
short-term denaturation	0.120 ± 0.010		
ratio PPI/Fab			
1		0.050 ± 0.003	0.040 ± 0.003
5		1.000 ± 0.005	0.053 ± 0.002
10		0.110 ± 0.005	0.063 ± 0.005
20		0.110 ± 0.008	0.069 ± 0.007
30		0.110 ± 0.008	0.064 ± 0.003
30 + 20 μM CsA		0.032 ± 0.003	
Maximum yield (%)	30	39	40

The second example deals with the renaturation, purification and down-stream processing of antibody fragments. As with immunotoxin, cytoplasmic expression of murine antibody chains (MAK33) in *E. coli* results in the formation of inclusion bodies. Rudolph, Buchner and coworkers designed a procedure for renaturation which produces microbially expressed authentic Fab-fragments at levels of up to 40% of the total amount of recombinant protein. Faced with a system known to show 'assisted folding' in the cell, a whole set of solvent parameters (temperature, protein concentration, redox buffer, labelling components) had to be varied to mimic conditions *in vivo* (Buchner & Rudolph 1991). More recently, Lilie *et al.* (1993, 1994) included folding catalysts in the investigation. It turns out that in the case of the oxidized Fab fragment peptidyl prolyl isomerases (PPIs) not only accelerate the refolding reaction, but also increase the proportion of correctly folded molecules, thus proving that catalysis affects kinetic partitioning by enhancing a rate-limiting step on the folding path (see table 2). Obviously, proline *cis–trans* isomerization is involved in the folding reaction. However, apart from acting as a folding catalyst, PPI also stabilizes folding intermediates in a similar way to serum albumin, or to the increasing concentration of native protein in the pulse renaturation approach mentioned previously.

PDI has no chaperone effect on the renaturation of oxidized Fab. Instead, this enzyme increases the yield of reactivation, at the same time shifting the redox dependence from a GSH^2:GSSG ratio around 10 mM to less than 1 mM observed for the spontaneous reaction. Again, there is kinetic competition but this time it is between domain folding and the interaction of PDI with its target cysteine residues (Lilie *et al.* 1994).

(c) Reverse micelles and immobilized proteins

Examining the previous results, it could be assumed that overexpression of a specific protein alongside chaperones and folding catalysts in the same plasmid would finally yield 100% of the desired protein in its native state. So far, active research toward this goal has shown little success. It may be worthwhile, therefore, to devise other means by which off-pathway reactions may be eliminated. There are two alternatives for consideration: (i) reconstitution in reverse micelles; and (ii) folding of polypeptide chains bound to solid matrices.

Regarding the properties of proteins in reverse micelles, it has been found that at low water content (< 6%) proteins show enhanced stability (Luisi & Magid 1986; Gómez-Puyou 1992). Mixing a denatured protein (e.g. in 6 M guanidinium chloride) with micelle-forming compounds, at a sufficiently low protein:micelle ratio, a more or less monodisperse system may be established in which at most one monomer per micelle is present (Luisi & Magid 1986, p.447 f). Because the micelles contain renaturation buffer, isolated 'caged' subunits will form structured monomers in separate micelles. Only when the micelles merge or protein transfer is complete, can assembly occur. By using this approach, preliminary reconstitution experiments with oligomeric enzymes resulted in high levels of reactivation without significant side reactions (A. Gómez-Puyou *et al.*, unpublished results; R. Jaenicke, unpublished results).

The idea of using matrix-bound polypeptides in protein folding goes back to experiments designed to determine the catalytic properties of isolated subunits of oligomeric enzymes. By using low levels of activation and low protein concentrations, Chan (1970) succeeded in fixing oligomers only at one or very few sites so that after denaturation–dissociation and subsequent washing, monomers only were covalently bound to the matrix. Their renaturation led to unexpectedly high yields which could easily be quantitated by hybridization and similar methods.

Expanding this idea, more recent folding studies on immobilized enzymes have shown that the properties of the matrix (solid or gel, cross-linking, porosity, polarity, etc.) and the linker connecting the protein to the matrix need careful consideration (Gottschalk & Jaenicke 1991). It is obvious that this approach will be adapted for technological application; using poly-

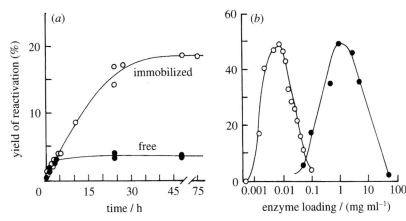

Figure 2. Renaturation–reactivation of soluble and immobilized. α-glucosidase-arg$_6$ after denaturation in 6M GdmCl. (*a*) Kinetics of reactivation at 0.1 mg ml^{-1} α-glucosidase concentration. (*b*) Profiles of the kinetic competition of folding and aggregation. The immobilized enzyme allows reconstitution up to 5 mg ml^{-1} (G. Stempfer & R. Rudolph, unpublished data).

ionic N- or C-terminal tails to immobilize the protein to an ion-exchange resin in a reversible fashion. The outcome in the case of α-glucosidase is most encouraging: the enzyme with its arg$_6$ tail shows long-term stability over a period of weeks; its reactivation yield after preceding GdmCl denaturation is increased at least fivefold, and the upper limit of protein concentration where maximum reactivation without aggregation can still be accomplished, is shifted from 10 μg ml^{-1} to 5 mg ml^{-1} (see figure 2).

6. KINETIC PARTITIONING: CELLULAR ASPECTS

There are three possible rate-limiting steps in the self organization of proteins: disulphide formation; proline *cis–trans* isomerization; and self assembly. On the other hand, there are two enzymes, localized in the appropriate cellular compartments, which catalyse the first two steps, and a whole range of chaperones and helper proteins to assist the third reaction.

Faced with the wealth of recent data which seem to round up in well-defined reaction cycles, one might suspect that, regarding the 'vision' of how kinetic partitioning *in vivo* might work, it is sufficient to summarize briefly the enzyme mechanisms and specificities, on one hand, and the regulation of intra- and intermolecular interactions in the cavities of chaperones, on the other. However, both folding catalysis and chaperoning are far from being solved. The current state of our knowledge is reviewed in Jaenicke & Buchner (1993); only certain new aspects will be outlined here.

(a) Folding catalysts

The biological significance of PDI, PPI and chaperones is now well-established. The three-dimensional structure of representatives of all three types of accessory protein have been resolved. DsbA, the PDI-homologue in *E. coli*, shows a close relation to thioredoxin; its second domain may be involved in communication with a more complex mechinery, including a membrane-spanning unit (DsbB). The anomalous redox and stability properties of DsbA (with the reduced form of the enzyme showing higher stability than the oxidized one) have been resolved elegantly by a number of recent studies (Bardwell & Beckwith 1993; Wunderlich & Glockshuber 1993).

It seems that PPIs are ubiquitous enzymes, catalysing rotation around the X-pro peptide bond (at least *in vitro*) and inhibiting the signal transduction processes involved in, for example, immuno-suppression. The three-dimensional structure of two different representatives of the PPI family has been elucidated, supporting a 'catalysis by distortion' mechanism. The substrate specificity is low.

Obviously, PPIs serve various functions in the cell. To date there is little direct evidence that they serve as folding catalysts, apart from the location of the enzyme in the endoplasmic reticulum, and its effect on the formation of collagen in fibroblasts.

(b) Chaperones

With respect to the involvement of chaperones in the folding and targeting of proteins, the field is moving so fast (and producing many inconsistent results) that it seems premature to propose general mechanisms of chaperone action and interaction even for the best-known systems DnaK–DnaJ–GrpE and GroEL–GroES (Jaenicke 1993; Jaenicke & Buchner 1993). X-ray structures are now becoming available and therefore cartoons should soon become obsolete.

Open questions about the role of GroEL and GroES deal with the stoichiometries and roles of the various components in the cycle of ATP hydrolysis, the substrate specificity and the conformational state of the bound polypeptide and, finally, the functional differences of single- versus double-doughnut assemblies, as well as symmetrical versus asymmetrical quaternary structures. Some of these questions have been addressed in a series of recent papers (Azem *et al.* 1994; Schmidt *et al.* 1994; Todd *et al.* 1994) which suggest that GroEL–GroES exhibit half-of-the-sites activity, involving both asymmetrical and symmetrical forms of the complex. As a result of a single turnover, unfolded protein (originally bound to the asymmetrical complex) dissociates in a non-native state, in a way that is consistent with intermolecular transfer of the substrate protein between toroids of high and low affinity. It remains to be shown how this model can be correlated or combined with other reaction schemes proposed on the basis of equilibrium experiments (cf. Hendrick & Hartl 1993).

Whether chaperones catalyse protein folding, acting as helper proteins and 'unfoldases', is still controversial. In numerous cases chaperones have been shown to slow down reactivation, probably due to their high affinity for their unfolded substrate. 'Resurrection of aggregates' is the exception rather than the rule: for steric reasons because they do not fit into the 'active sites'; or for energetic reasons because aggregates are commonly trapped in a deep energy well. Support for the steric argument comes from lag phases observed in the GroE-assisted formation of oligomeric proteins (Jaenicke & Buchner 1993).

Regarding the state of association of the released protein substrate, direct proof for the monomeric state (i.e. against oligomerization on the chaperone) came from reconstitution experiments with glutamin synthetase, where the formation of the native dodecamer has been shown to occur in solution after release of the subunits from the binary GroEL-complex (Fisher 1993). What drives the release, and what the polypepticle chain in the chaperone-substrate complex looks like, is still unresolved.

With respect to protection of the nascent polypeptide chain from misfolding, or chemical modification, the question arises 'when in the life of a protein, folding and chaperoning start'. Experimental examination of this question is difficult because of the heterogeneity and low concentration of nascent polypeptides. Previous work pointed to cotranslational folding-by-parts. Recently, monoclonal antibodies have been used to show that antigen recognition may already start at the subdomain level (Friguet *et al.* 1994).

Subdomains are approximately twice the size required by DnaJ for the recognition of its substrate. How the binding of the DnaJ–Hsp 70 chaperone assembly correlates with the *in vitro* translation studies remains to be shown. It seems there may be still earlier processes involved. It has been established for a long time that in the case of proteins targeted to the endoplasmic reticulum, complex formation with SRP starts when the polypeptide is just leaving the ribosome (Rapoport 1992). There is one more component, NAC, involved in targeting, which raises the question does the nascent chain leave the ribosome through a cleft rather than a channel? NAC is a heterodimeric protein which binds exclusively to the nascent polypeptide, shielding non-signal peptide regions from promiscuous interactions with the signal recognition particle. As shown by Wiedmann *et al.* (1994), in the absence of NAC, proteins lacking the signal peptide are recognized and mistargeted by the SRP. Adding NAC restores the specificity of SRP binding, as well as correct targeting and translocation. Obviously, NAC binds to signal peptides that are only partly exposed from the ribosome, serving as an adaptor between the translation complex and the cellular folding and transport machineries, thereby protecting nascent chains from premature and inappropriate interactions. Thus, the nascent chain is tunnelled through a groove between the ribosome and NAC. Correspondingly, the length of the amino-acid sequence accessible to the cytosol is unexpectedly short: using truncated mRNA as a means to produce fixed peptides of constant length, it turns out that no more than *ca.* 15 amino-acid residues away from the ribosomal peptidyl transferase site the growing polypeptide becomes the target of chaperone-like components regulating their interactions (M. Wiedmann, personal communication).

In summary, the picture that emerges from studies of both prokaryotes and eukaryotes clearly shows that off-pathway reactions in the cell are blocked by a whole arsenal of components from the moment the peptide emerges. The reason why, in spite of this protection, protein biosynthesis and subsequent structure formation commonly does not yield 100 % (cf. Hurtley & Helenius 1989) seems obvious: interactions involved in chaperone action must not be too strong, otherwise the nascent polypeptide chains would be trapped. Thus, during the off-reaction of complex formation, molecules may escape from the folding pathway ending up in protein turnover.

This review was written during a stay at the Fogarty International Center, NIH, Bethesda, Maryland. Fruitful discussions with Dr J. Bardwell, Dr J. Buchner, Dr R. Glockshuber, Dr F.-U. Hartl, Dr G. Lorimer, Dr R. Seckler and Dr M. Wiedmann are gratefully acknowledged. Work carried out in the author's laboratory was generously supported by grants from the Deutsche Forschungsgemeinschaft, the Fonds der Chemischen Industrie, and Boehringer Mannheim GmbH.

This paper is dedicated to Professor Hans Neurath on the occasion of his 85th birthday.

REFERENCES

Azem, A., Kessel, M. & Goloubinoff, P. 1994 Characterization of a functional $GroEL_{14}$ $(GroES_7)_2$ chaperonin hetero-oligomer. *Nature, Lond.* **265**, 653–656.

Baldwin, R.L. 1991 Experimental studies of pathways of protein folding. *CIBA Found. Symp.* **161**, 190–205.

Bardwell, J. & Beckwith, J. 1994 The bonds that tie: catalyzed disulfide bond formation. *Cell* **74**, 769–771.

Blond-Elguindi, S. & Goldberg, M.E. 1990 Kinetic characterization of early immunoreactive intermediates during refolding. *Biochemistry* **29**, 2409–2417.

Bowden, G.A., Paredes, A.M. & Georgiou, G. 1991 Structure and morphology of inclusion bodies in E. coli. *Biotechnology* **9**, 725–730.

Brems, D.N., Plaisted, S.M., Kaufmann, E.W., Lund, M. & Tomich, C.-S.C. 1988 Stabilization of an associated folding intermediate of BGH by site-directed mutagenesis. *Proc. natn. Acad. Sci. U.S.A.* **85**, 3367–3371.

Buchner, J. & Rudolph, R. 1991 Renaturation, purification and characterization of $recF_{ab}$-fragments produced in *E. coli. Biotechnology* **9**, 157–162.

Buchner, J., Brinkmann, U. & Pastan, I. 1992 Renaturation of a single-chain immunotoxin facilitated by chaperones and PDI. *Biotechnology* **10**, 682–685.

Chan, W.W.-C. 1970 Matrix-bound protein subunits. *Biochem. biophys. Res. Commun.* **41**, 1198–1204.

Danner, M. & Seckler, R. 1993 Mechanism of phage P22 tailspike protein folding mutation. *Protein Sci.* **2**, 1869–1881.

Fisher, M.T. 1993 On the assembly of dodecameric glutamine synthetase from stable chaperonin complexes. *J. biol. Chem.* **268**, 13777–13779.

Friguet, B., Djavady-Ohaniance, L., King, J. & Goldberg, M.E. 1994 *In vitro* and ribosome- bound folding intermediates of P22 tailspike protein detected with monoclonal antibodies. *J. biol. Chem.* **269**, 15945–15949.

Fuchs, A., Seiderer, C. & Seckler, R. 1991 *In vitro* folding of phage P22 tailspike protein. *Biochemistry* **30**, 6598–6604.

Goldberg, M.E., Rudolph, R. & Jaenicke, R. 1991 A kinetic study of the competition between renaturation and aggregation of lysozyme. *Biochemistry* **30**, 2790–2797.

Gómez-Puyou, A. (ed.) 1992 *Biomolacules in organic solvents.* Boca Raton: CRC Press.

Gottschalk, N. & Jaenicke, R. 1991 Authenticity and reconstitution of immobilized enzymes. *Biotech. Appl. Biochem.* **14**, 324–335.

Helenius, A., Marquardt, T. & Braakman, I. 1992 The endoplasmic reticulum as a protein-folding compartment. *Trends Cell Biol.* **2**, 227–231.

Hendrick, J.P. & Hartl, F.-U. 1993 Molecular chaperone functions of heat-shock proteins. *A. Rev. Biochem.* **62**, 349–384.

Hurtley, S.M. & Helenius, A. 1989 Protein oligomerization in the ER. *A. Rev. cell Biol.* **5**, 277–307

Jaenicke, R. 1967 Intermolecular forces in the process of heat aggregation of globular proteins. *J. Polymer Sci.* **16**, 2143–2160.

Jaenicke, R. 1987 Folding and association of proteins. *Progr. Biophys. molec. Biol.* **49**, 117–237.

Jaenicke, R. & Rudolph, R. 1988 Folding proteins. In *Protein structure: a practical approach* (ed. T.E. Creighton), pp. 191–223. Oxford: IRL Press.

Jaenicke, R. 1991a Protein stability and molecular adaptation to extreme conditions. *Eur. J. Biochem.* **202**, 715–728.

Jaenicke, R. 1991b Protein folding: Local structures, domains and assemblies. *Biochemistry* **30**, 3147–3161.

Jaenicke, R. 1993a What does protein refolding *in vitro* tell us about protein folding in the cell? *Phil. Trans. R. Soc.Lond.* B **339**, 287–295.

Jaenicke, R. 1993b Role of accessory proteins in protein folding. *Curr. Opin. struct. Biol.* **3**, 104–112.

Jaenicke, R. & Buchner, J. 1993 Protein folding: From 'unboiling an egg' to 'catalysis of folding'. *Chemtracts: Biochem. molec. Biol.* **4**, 1–30.

Jaenicke, R. 1994 Eye lens proteins: Structure, super-structure, stability and genetics. *Naturwissenschaften.* **81**, 423–429.

Jones, C.M., Henry, E.R., Hu, Y., Chan, C.-K., Luck, S.D., Bhuyan, A., Roder, H., Hofrichter, J. & Eaton, W.A. 1993 Fast events in protein folding initiated by ns laser photolysis. *Proc. natn. Acad. Sci. U.S.A.* **90**, 11860–11864.

Kiefhaber, T., Rudolph, R., Kohler, H.-H. & Buchner, J. 1991 Protein aggregation *in vitro* and *in vivo*. *Biotechnology* **9**, 825–829.

Lilie, H., Lang, K., Rudolph, R. & Buchner, J. 1993 Prolin isomerases catalyze antibody folding *in vitro*. *Protein Sci.* **2**, 1490–1496.

Lilie, H., McLaughlin, S., Freedman, R. & Buchner, J. 1994 Influence of PDI on antibody folding in vitro. *J. biol. Chem.* **269**, 14290–14296.

Luisi, P.L. & Magid, L.J. 1986 Solubilization of enzymes and nucleic acids in hydrocarbon micellar solutions. *CRC Crit. Rev. Biochem.* **20**, 409–474.

Mayr, E.-M., Jaenicke, R. & Glockshuber, R. 1994 Domain interactions and connecting peptides in lens crystallins. *J. molec. Biol.* **235**, 84–88.

Mitraki, A. & King, J. 1989 Protein folding intermediates and inclusion body formation. *Biotechnology* **7**, 690–697.

Mitraki, A. & King, J. 1992 Amino acid substitutions influencing intracellular protein pathways. *FEBS Lett.* **307**, 20–25.

Mitraki, A., Danner, M., King, J. & Seckler, R. 1993 Ts-mutations and second-site suppressor substitutions affect folding of the P22 tsp *in vitro*. *J. biol. Chem.* **268**, 20071–20075.

Opitz, U., Rudolph, R., Jaenicke, R., Ericsson, L. & Neurath, H. 1987 Proteolytic dimers of porcine LDH. *Biochemistry* **26**, 1399–1406.

Rapoport, T.A. 1992 Transport of proteins across the ER membrane. *Science, Wash.* **258**, 931–936.

Rudolph, R. 1990 Renaturation of recombinant, disulfide-bonded proteins from 'inclusion bodies'. In *Modern methods in protein and nucleic acid research* (ed. H. Tschesche), pp. 149–171. Berlin, New York: de Gruyter.

Schmidt, M., Rutkat, K., Rachel, R., Pfeifer, G., Jaenicke, R., Viitanen, P., Lorimer, G. & Buchner, J. 1994 Symmetric complexes of GroE chaperonins as part of the protein folding cycle. *Science, Wash.* **265**, 656–659

Tandon, S. & Horowitz, P. 1986 Detergent-assisted refolding of GdmCl-denatured rhodanese. *J. biol. Chem.* **261**, 15615–15618.

Teschner, W., Rudolph, R. & Garel, J.-R. 1987 Intermediates on the folding pathway of ODH from *Pecten jacobaeus*. *Biochemistry* **26**, 2791–2796.

Todd, M.J., Viitanen, P.V. & Lorimer, G.H. 1994 Dynamics of the chaperonin ATPase cycle: Implications for facilitated protein folding. *Science, Wash.* **265**, 659–666.

Wiedmann, B., Sakai, H., Davis, T.A. & Wiedmann, M. 1994 A protein complex required for signal-sequence-specific sorting and translocation. *Nature, Lond.* **370**, 434–440.

Wunderlich, M. & Glockshuber, R. 1993 Redox properties of PDI (DsbA) from *E. coli*. *Protein Sci.* **2**, 717–726.

Principles of chaperone-mediated protein folding

F. ULRICH HARTL

Howard Hughes Medical Institute and Cellular Biochemistry and Biophysics Program, Memorial Sloan-Kettering Cancer Center, 1275 York Avenue, New York, New York 10021, U.S.A.

SUMMARY

The recent discovery of molecular chaperones and their functions has changed dramatically our view of the processes underlying the folding of proteins *in vivo*. Rather than folding spontaneously, most newly synthesized polypeptide chains seem to acquire their native conformations in a reaction mediated by chaperone proteins. Different classes of molecular chaperones, such as the members of the Hsp70 and Hsp60 families of heat-shock proteins, cooperate in a coordinated pathway of cellular protein folding.

1. INTRODUCTION

In vitro, many unfolded proteins are able to fold to their native conformations spontaneously. This observation, first made by Anfinsen about three decades ago, demonstrated that all the information necessary to specify the three-dimensional structure of a protein is contained in its linear amino acid sequence (Anfinsen 1973). Consequently, it had been assumed that the *de novo* folding of proteins upon synthesis on ribosomes also generally occurs spontaneously. This view has changed profoundly over the past six years, due to the discovery of a large number of proteins, known as 'molecular chaperones', which are essential for cellular protein folding and occur ubiquitously in all types of cell in the cytosol as well as in various subcellular membrane compartments (Ellis 1987; Hartl *et al.* 1994).

The molecular chaperone concept

The term 'molecular chaperone' was coined for nucleoplasmin, a protein that binds to histones and mediates nucleosome assembly (Laskey *et al.* 1978). Molecular chaperone proteins of several structurally unrelated classes, many of them stress or heat-shock proteins, are now known to participate in a variety of cell functions. They facilitate *de novo* protein folding under normal growth conditions, prevent protein aggregation under stress conditions and stabilize polypeptide chains in an unfolded state for translocation across organellar membranes (Hendrick & Hartl 1993; Ellis 1994; Hartl *et al.* 1994; Stuart *et al.* 1994). Several lines of cell biological research contributed to the formulation of the novel concept of assisted protein folding: the isolation of not-yet-assembled subunits of ribulose bisphosphate carboxylase oxygenase (Rubisco) in chloroplasts as a complex with a high molecular mass binding-protein, the Rubisco subunit binding protein (RSBP), suggested a critical role of this component in Rubisco assembly (Barraclough & Ellis 1980). RSBP was later found to

be the chloroplast homolgue of *E. coli* GroEL and mitochondrial Hsp60, which have been classified as members of the Hsp60 or 'chaperonin' family of molecular chaperones (Hemmingsen *et al.* 1988; McMullin & Hallberg 1988). Mutations in the genes encoding GroEL and its co-factor GroES had been reported to affect the assembly of bacteriophage particles (Georgopoulos *et al.* 1973). The oligomeric assembly of proteins imported into mitochondria from the cytosol was found to be defective in a yeast strain containing a mutated version of mitochondrial Hsp60 (Cheng *et al.* 1988). The primary function of Hsp60 was subsequently shown to be to mediate the folding of monomeric polypeptide chains (Ostermann *et al.* 1989) and the subunits of oligomeric proteins in an ATP-dependent reaction (Zheng *et al.* 1993).

As an independent line of evidence, the Hsp70s, another major class of molecular chaperones, were proposed to protect certain proteins from denaturation under heat-stress (Pelham 1986) and were shown to associate with ribosome-bound polypeptides (Chirico *et al.* 1988; Deshaies *et al.* 1988; Beckmann *et al.* 1990; Frydman *et al.* 1994). A common theme in all these studies was that the binding proteins stabilized the otherwise unstable conformations of non-native proteins which are prone to aggregation. It is now generally believed that molecular chaperones shield the hydrophobic sequences or surfaces exposed by conformational intermediates on the protein folding pathway. They do not recognize a consensus sequence motif and therefore have the ability to prevent the incorrect intra- and intermolecular folding and association of many different proteins. The Hsp70s and Hsp60s then promote correct folding by repeatedly binding and releasing their substrate proteins regulated by ATP binding and hydrolysis. In this process the molecular chaperones do not typically function as catalysts of protein folding. Generally, they increase the yield of a folding reaction rather than its speed. Once a protein has reached its native state, it no longer presents hydrophobic surfaces for chaperone binding. However, exposure to certain forms of cellular stress,

Phil. Trans. R. Soc. Lond. B (1995) **348**, 107–112
Printed in Great Britain

107

such as heat-stress, may cause the partial or complete unfolding of proteins, leading to their renewed interaction with chaperones.

2. CELLULAR FOLDING PATHWAYS

The function of the Hsp70s and Hsp60s in protein folding is understood in considerable detail. Both groups of chaperones play an essential role in the folding of newly synthesized polypeptides in the cytosol, as well as in mitochondria and chloroplasts (Hendrick & Hartl 1993; Hartl *et al.* 1994). Here they cooperate in what appears to be a general pathway of cellular protein folding (figure 1) (Langer *et al.* 1992; Frydman *et al.* 1994; Stuart *et al.* 1994). The Hsp70s interact with the nascent polypeptide at a very early stage of chain elongation. Given the high density of total protein in the cytosol (20–30%) and of unfolded protein molecules (up to 30–50 μM in *E. coli*), this interaction appears to be necessary to prevent the

aggregation of nascent chains or their unfavourable association with the ribosome surface at a point when the partially synthesized and therefore conformationally restricted polypeptide is not yet able to form a stable tertiary structure. At a later stage, the not yet folded polypeptide can be transferred to a chaperonin of the Hsp60 family (in bacteria, mitochondria and chloroplasts) or the TCP-1 family (in the eukaryotic cytosol), which mediates folding to the native state (Langer *et al.* 1992; Frydman *et al.* 1994). In the case of the TCP-1 ring complex (TRiC), this transfer can occur co-translationally before completion of synthesis (Frydman *et al.* 1994).

(a) Mechanism of the Hsp70 system

The Hsp70s have the ability to bind short, extended peptide segments of seven or eight residues which are enriched in hydrophobic amino acids (Flynn *et al.* 1991; Blond-Elguindi *et al.* 1993). Only more recently has it been realized that for full function the Hsp70s

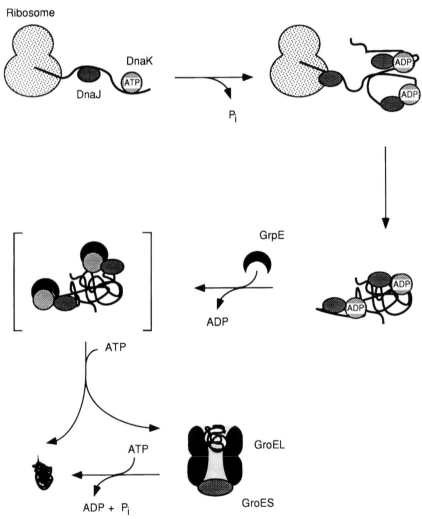

Figure 1. Model for the pathway of protein folding in the *E. coli* cytosol. The polypeptide chain emerging from the ribosome is bound by DnaJ and DnaK (Hsp70) (Hendrick *et al.* 1993; Gaitanaris *et al.* 1994). The direct interaction between DnaK and DnaJ in the presence of ATP leads to the formation of a ternary complex between nascent chain, DnaK and DnaJ in which DnaK is in the ADP-state. This complex dissociates upon the GrpE-dependent dissociation of ADP and the binding (not hydrolysis) of ATP to DnaK (Szabo *et al.* 1994). The protein may then fold to the native state by multiple rounds of interaction with the DnaK, DnaJ, GrpE system or is transferred for final folding to GroEL/GroES (Langer *et al.* 1992*a*). We assume that a large fraction of cytosolic proteins have to interact with both chaperone systems to reach the native state.

depend on the regulation by further proteins. In *E. coli* the Hsp70, called DnaK, cooperates with the chaperone DnaJ and the nucleotide exchange factor GrpE, proteins of about 43 kDa and 23 kDa, respectively (Georgopoulos 1992). Both DnaK and DnaJ bind nascent polypeptide chains cotranslationally (Hendrick *et al.* 1993; Gaitanaris *et al.* 1994; Kudlicki *et al.* 1994), whereby DnaJ seems to mediate the loading of DnaK onto the elongating chain (figure 1). The interaction of DnaJ with DnaK accelerates the hydrolysis of DnaK-bound ATP to ADP (Liberek *et al.* 1991) and stabilizes the ADP-state of DnaK which has a high affinity for unfolded polypeptide (Palleros *et al.* 1993). As a result, a stable ternary complex consisting of polypeptide substrate, DnaJ and ADP-bound DnaK is formed (Langer *et al.* 1992; Szabo *et al.* 1994). GrpE then functions as a nucleotide exchange factor for DnaK in dissociating the bound ADP, whereupon ATP binding to DnaK causes the release of the polypeptide substrate (Szabo *et al.*, 1994). This allows the transfer of the unfolded protein to GroEL, the bacterial Hsp60 (Langer *et al.* 1992). At least *in vitro*, the folding of certain proteins may be achieved through ATP-dependent cycles of binding and release to DnaK and DnaJ alone (Schröder *et al.*, 1993; Szabo *et al.*, 1994). *In vivo*, however, the primary function of the Hsp70 system seems to be in maintaining the polypeptide chain in a non-aggregated state, competent for folding by the chaperonin. The eukaryotic cytsosol contains several DnaJ homologues (Caplan & Douglas 1993), but a structural or functional equivalent of GrpE has not yet been identified in this compartment.

(b) *Mechanism of the chaperonin system*

While the Hsp70 and DnaJ proteins function as monomers or dimers, the chaperonin of *E. coli*, GroEL, is a large complex consisting of two stacked rings of seven identical 60 kDa subunits, forming a cylinder with a central cavity (Hendrix 1979; Hohn *et al.* 1979; Langer *et al.* 1992*b*; Saibil *et al.* 1993; Braig *et al.* 1994). GroEL has an essential cofactor, GroES, a single heptameric ring of 10 kDa subunits that binds to GroEL and increases the cooperativity of ATP hydrolysis in the GroEL ring system. Although this regulation is not required for the ATP-dependent release of bound protein from GroEL per se, with many substrate proteins it is necessary to make the release reaction productive for folding. Under most conditions, binding of GroES to either end of the GroEL cylinder strongly reduces the affinity of the opposite end for binding a second GroES (Langer *et al.* 1992; Saibil *et al.* 1993; Chen *et al.* 1994). This negative cooperativity of GroES binding is decreased at high concentrations of Mg^{2+} (15–50 mM) and at elevated pH (pH 7.7–8.0), conditions which allow the formation of symmetrical GroES:GroEL:GroES complexes (Llorca *et al.* 1994; Schmidt *et al.* 1994). A recent kinetic analysis of the GroEL–GroES reaction cycle using the new technique of surface plasmon resonance (Biacore™) failed to demonstrate the functional significance of these so-called 'football' structures (M.K. Hayer-Hartl, unpublished results).

The interaction between GroEL and GroES is dynamic, both in the absence and presence of substrate polypeptide. The asymmetrical binding of GroES stabilizes the seven subunits of GroEL that are in contact with GroES in a tight ADP state, resulting in a 50% inhibition of the GroEL ATPase (Martin *et al.* 1993; Todd *et al.* 1993). Binding and hydrolysis of ATP in the opposite heptamer of GroEL causes the transient release of the tightly bound ADP and GroES. This cycling of GroES between bound and free states is normally slow but is accelerated by the association of polypeptide substrate with GroEL (Martin *et al.* 1993). Polypeptide binding stimulates the ATPase activity of GroEL (Martin *et al.* 1991; Jackson *et al.* 1993). The unfolded polypeptide binds initially to the GroEL ring that is not covered by GroES (figure 2). This facilitates the release of the tightly bound ADP and the dissociation of GroES. Upon ATP binding, GroES may then reassociate with the protein-containing ring of GroEL, inducing the ATP-hydrolysis-dependent release of the bound polypeptide (Martin *et al.* 1993). Interestingly, binding of GroES causes a massive outwards movement of the apical domains of the GroEL subunits, creating an enclosed, dome-shaped space with a maximum height and width of 70 Å (figure 2) (Chen *et al.* 1994). GroES could initially make contact with the outer surface of the GroEL cylinder, triggering further domain movement, thus transiently displacing the polypeptide substrate into the cavity for folding (Fenton *et al.* 1994; Hartl 1994). At least partial folding may thus occur in a shielded microenvironment, before the polypeptide emerges from the chaperonin cavity (Martin *et al.* 1993). Multiple rounds of binding and release to GroEL may be necessary for completion of folding (Martin *et al.* 1991). Alternatively, GroES may exert its function aiding productive protein release from the GroEL ring that is not occupied by the folding polypeptide. Both mechanisms of GroES action may not be mutually exclusive (figure 2).

GroEL binds its substrate in the conformation of a compact, yet flexible, molten globule-state which exposes hydrophobic surfaces to solvent (Martin *et al.* 1991; Hayer-Hartl *et al.* 1994; Robinson *et al.* 1994). The structure-based mutational analysis of GroEL indeed suggests the presence of a complementary hydrophobic binding surface that lines the cavity of the cylinder (Fenton *et al.* 1994). In contrast to Hsp70, GroEL does not seem to recognize short peptide sequences in extended conformations (Landry *et al.* 1992).

The exact extent of folding that can occur while a polypeptide is in association with the chaperonin, either bound to its surface or upon release into its cavity, remains to be defined. Evidence has been presented that the substrate polypeptide is released into the bulk solution in a conformation that is significantly less prone to aggregation than the conformation initially bound by the chaperonin (Martin *et al.* 1991). While small, single-domain proteins, such as barnase, may reach their native state in association with GroEL (Gray & Fersht 1993), other proteins can be released before they have reached the native state

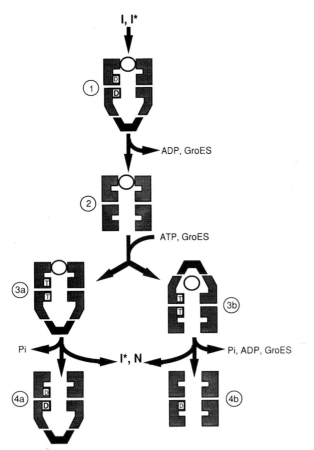

Figure 2. Model for the ATP-dependent interaction between GroEL, GroES and folding polypeptide. D (bold), the high-affinity ADP state in the seven subunits of GroEL which are bound to GroES; D (not bold), the lower ADP affinity of the subunits in the opposite toroid which may hydrolyse ATP (Martin *et al.* 1993); T, the subunits in a GroEL toroid in the ATP-bound state; I, polypeptide substrate as compact folding intermediate; I*, folding intermediate part way advanced towards the native state (for polypeptides with separate domains or subdomains); N, native protein. 1, Polypeptide binding facilitates dissociation of tightly bound ADP and GroES. 2, Polypeptide transiently bound in nucleotide-free toroid. 3, ATP and GroES rebind. GroES associates either with free GroEL ring (3a) or with polypeptide-containing ring (3b). ATP-hydrolysis leads to polypeptide release (4a) and incompletely folded polypeptide rebinds in (1). In 3b, polypeptide is transiently enclosed in the central cavity and is free to fold. ATP-hydrolysis in the GroES-bound toroid then generates the tight ADP-state (not shown). ADP dissociates upon ATP-hydrolysis in the opposite ring, causing dissociation of GroES and allowing polypeptide release. Polypeptide may rebind in (2).

and subsequently re-bind to the chaperonin (Martin *et al.* 1991; Weissman *et al.* 1994).

3. PERSPECTIVES

The functional cooperation between molecular chaperones in protein folding is emerging as a common theme from a number of recent studies (Hendrick & Hartl 1993; Hartl *et al.* 1994). This applies also to the endoplasmic reticulum, the compartment that is responsible for the folding and assembly of secretory proteins. During translocation across the endoplasmic reticulum (ER) membrane, secretory proteins probably interact first with the Hsp70 homologue BiP, followed by interactions with various chaperones (Helenius *et al.* 1992), including the membrane-bound protein calnexin (Bergeron *et al.* 1994). Unlike the cytosol, the environment of the ER lumen is oxidizing and contains the enzyme protein disulphide isomerase which accelerates the correct formation of disulphide bonds (Freedman 1989). A ring-shaped chaperonin is apparently absent from the ER.

With respect to the protein folding problem, traditionally the domain of biophysicists and theoreticians, it will be interesting to see whether molecular chaperones may be able to influence the pathways of protein folding or even the final outcome of a folding reaction. Are there situations where the information specified in the linear sequence of amino acid residues is not sufficient for folding to the native state? Has the coevolution of proteins and chaperones perhaps favoured certain folding pathways over others? To address these questions, the conformational dynamics of chaperone-substrate protein interactions will have to be resolved.

Another important direction of research will be the analysis of protein folding in the context with translation. Very little is known about the very early events of folding that may occur when the growing polypeptide chain is still within the ribosomal exit tunnel or groove. To what extent does the formation of secondary and tertiary structure proceed co-translationally? Here the final goal would be the *in vitro* reconstitution of translation and folding of a nascent polypeptide chain with all the necessary components in purified form.

Reserach in the author's laboratory is supported by the National Institutes of Health and by the Howard Hughes Medical Institute.

REFERENCES

Anfinsen, C.B. 1973 Principles that govern the folding of protein chains. *Science, Wash.* **181**, 223–230.

Barraclough, R. & Ellis, R.J. 1980 Protein synthesis in chloroplasts. IX. Assembly of newly synthesized large subunits into ribulose bisphosphate carboxylase in isolated intact pea chloroplasts. *Biochim. biophys. Acta* **608**, 18–31.

Beckmann, R.P., Mizzen, L. & Welch, W. 1990 Interaction of Hsp70 with newly synthesized proteins: implications for protein folding and assembly. *Science, Wash.* **248**, 850–856.

Bergeron, J.J.M., Brenner, M.B., Thomas, D.Y. & Williams, D.B. 1994 Calnexin: a membrane-bound chaperone of the endoplasmic reticulum. *Trends Biochem. Sci.* **19**, 124–128.

Blond-Elguindi, S., Cwirla, S.E., Dower, W.J., Lipshutz, R.J., Sprang, S.R., Sambrook, J.F. & Gething, M.-J.H. 1993 Affinity panning of a library of peptides displayed on bacteriophages reveals the binding specificity of BiP. *Cell* **75**, 717–728.

Braig, K., Furuya, F., Hainfeld, J. & Horwich, A.L. 1993 Gold-labeled DHFR binds in the center of GroEL. *Proc. natn. Acad. Sci. U.S.A.* **90**, 3978–3982.

Braig, K., Otwinowski, Z., Hegda, R., Boisvert, D., Joahimiak, A., Horwich, A.L. & Sigler, P.B. 1994 The crystal structure of GroEL at 2.8 Å resolution. *Nature, Lond.* **371**, 578–586.

Buchner, J., Schmidt, M., Fuchs, M., Jaenicke, R., Rudolph, R., Schmid, F.X. & Kiefhaber, T. 1991 GroE facilitates refolding of citrate synthase by suppressing aggregation. *Biochemistry* **30**, 1586–1591.

Caplan, A.J., Cyr, D. & Douglas, M.G. 1993 Eukaryotic homologues of *Escherichia coli* dnaJ: a diverse protein family that functions with HSP70 stress proteins. *Molec. Biol. Cell* **4**, 555–563.

Chen, S., Roseman, A.M., Hunter, A.S., Wood, S.P., Burston, S.G., Ranson, N.A., Clarke, A.R. & Saibil, H.R. 1994 Location of a folding protein and shape changes in GroEL–GroES complexes imaged by cryo-electron microscopy. *Nature, Lond.* **371**, 261–264.

Cheng, M.Y., Hartl, F.U., Martin, J., Pollock, R.A., Kalousek, F., Neupert, W., Hallberg, E.M., Hallberg, R.L. & Horwich, A.L. 1989 Mitochondrial heat-shock protein hsp60 is essential for assembly of proteins imported into yeast mitochondria. *Nature, Lond.* **337**, 620–625.

Chirico, W.J., Waters, M.G. & Blobel, G. 1988 70K heat shock related proteins stimulate protein translocation into microsomes. *Nature, Lond.* **332**, 805–810.

Deshaies, R.J., Koch, B.D., Werner-Washburne, M., Craig, E.A. & Schekman., R. 1988 A subfamily of stress proteins facilitates translocation of secretory and mitochondrial precursor polypeptides. *Nature, Lond.* **332**, 800–805.

Ellis, J. 1987 Proteins as molecular chaperones. *Nature, Lond.* **328**, 378–379.

Ellis, R.J. 1994 Roles of molecular chaperones in protein folding. *Curr. Opin. Struct. Biol.* **4**, 117–122.

Fenton, W.A., Kashi, Y., Furtak, K. & Horwich, A.L. 1994 Functional analysis of the chaperonin GroEL: identification of residues required for polypeptide binding and release. *Nature, Lond.* **371**, 614–619.

Flynn, G.C., Rohl, J., Flocco, M.T. & Rothman, J.E. 1991 Peptide-binding specificity of the molecular chaperone BiP. *Nature, Lond.* **353**, 726–730.

Freedman, R.B. 1989 Protein disulfide isomerase: multiple roles in the modification of nascent secretory proteins. *Cell* **57**, 1069–1072.

Frydman, J., Nimmesgern, E., Ohtsuka, K. & Hartl, F.U. 1994 Folding of nascent poypeptide chains in a high molecular mass assembly with molecular chaperones. *Nature, Lond.* **370**, 111–117.

Gaitanaris, G.A., Vysokanov, A., Hung, S.-Z., Gottesman, M. & Gragerov, A. 1994 *Escherichia coli* chaperones are associated with nascent polypeptide chains and promote the folding of λ repressor. *Molec. Microbiol.* **14**, 861–869.

Georgopoulos, C. 1992 The emergence of the chaperone machines. *Trends Biochem. Sci.* **17**, 295–299.

Georgopoulos, C., Hendrix, R.W., Casjens, S.R. & Kaiser, A.D. 1973 Host participation in bacteriophage lambda head assembly. *J. molec. Biol.* **76**, 45–60.

Gray, T.E. & Fersht, A.R. 1993 Refolding of barnase in the presence of GroE. *J. molec. Biol.* **232**, 1197–1207.

Hartl, F.U. 1994 Protein folding: secrets of a double-doughnut. *Nature, Lond.* **371**, 557–559.

Hartl, F.-U., Hlodan, R. & Langer, T. 1994 Molecular chaperones in protein folding: the art of avoiding sticky situations. *Trends Biochem. Sci.* **19**, 20–25.

Hayer-Hartl, M.K., Ewbank, J.J., Creighton, T.E. & Hartl, F.U. 1994 Conformational specificity of the chaperonin GroEL for the compact folding intermediates of α-lactalbumin. *EMBO J.* **13**, 3192–3202.

Helenius, A., Marquart, T. & Braakman, I. 1992 The endoplasmic reticulum as a protein-folding compartment. *Trends Cell Biol.* **2**, 227–231.

Hemmingsen, S.M., Woolford, C., van der Vies, S., Tilly, K., Dennis, D.T., Georgopoulos, C.P., Hendrix, R.W. & Ellis, R.J. 1988 Homologous plant and bacterial proteins chaperone oligomeric protein assembly. *Nature, Lond.* **333**, 330–334.

Hendrick, J.P. & Hartl, F.U. 1993 Molecular chaperone functions of heat-shock proteins. *A. Rev. Biochem.* **62**, 349–384.

Hendrick, J.P., Langer, T., Davis, T.A., Hartl, F.U. & Wiedmann, M. 1993 Control of folding and membrane translocation by binding of the chaperone DnaJ to nascent polypeptides. *Proc. natn. Acad. Sci. U.S.A.* **90**, 10216–10220.

Hendrix, R.W. 1979 Purification and properties of GroE, a host protein involved in bacteriophage assembly. *J. molec. Biol.* **129**, 375–392.

Hohn, T., Hohn, B., Engel, A. & Wurtz, M. 1979 Isolation and characterisation of the host protein GroE involved in bacteriophage lambda assembly. *J. molec. Biol.* **129**, 359–373.

Jackson, G.S., Staniforth, R.A., Halsall, D.J., Atkinson, T., Holbrook, J.J., Clarke, A.R. & Burston, S.G. 1993 Binding and hydrolysis of nucleotides in the chaperonin catalytic cycle: implications for the mechanism of assisted protein folding. *Biochemistry* **32**, 2554–2563.

Kudlicki, W., Odom, O.W., Kramer, G. & Hardesty, B. 1994 Activation and release of enzymatically inactive, full-length rhodanese that is bound to ribosomes as peptidyl-tRNA. *J. biol. Chem.* **269**, 16549–16553.

Landry, S.J., Jordan, R., McMacken, R. & Gierasch, L.M. 1992 Different conformations for the same polypeptide bound to chaperones DnaK and GroEL. *Nature, Lond.* **355**, 455–457.

Langer, T., Lu, C., Echols, H., Flanagan, J., Hayer, M.K. & Hartl, F.U. 1992 a Successive action of molecular chaperones DnaK (Hsp70), DnaJ and GroEL (Hsp60) along the pathway of assisted protein folding. *Nature, Lond.* **356**, 683–689.

Langer, T., Pfeifer, G., Martin, J., Baumeister, W. & Hartl, F.U. 1992 b Chaperonin-mediated protein folding: GroES binds to one end of the GroEL cylinder which accommodates the protein substrate within its central cavity. *EMBO J.* **11**, 4757–4765.

Laskey, R.A., Honda, B.M. & Finch, J.T. 1978 Nucleosomes are assembled by an acidic protein which binds histones and transfers them to DNA. *Nature, Lond.* **275**, 416–420.

Liberek, K., Marszalek, J., Ang, D., Georgopoulos, C. & Zylicz, M. 1991 *Escherichia coli* DnaJ and GrpE heat shock proteins jointly stimulate ATPase activity of DnaK. *Proc. natn. Acad. Sci. U.S.A.* **88**, 2874–2878.

Llorca, O., Marco, S., Carrascosa, J.L. & Valpuesta, J.M. 1994 The formation of symmetrical GroEL–GroES complexes in the presence of ATP. *FEBS Lett.* **345**, 181–186.

Martin, J., Langer, T., Boteva, R., Schramel, A., Horwich, A.L. & Hartl, F.-U. 1991 Chaperonin-mediated protein folding at the surface of groEL through a 'molten globule'-like intermediate. *Nature, Lond.* **352**, 36–42.

Martin, J., Langer, T., Boteva, R., Schramel, A., Horwich, A.L. & Hartl, F.U. 1991 Chaperonin-mediated protein folding at the surface of groEL through a 'molten globule'-like intermediate. *Nature, Lond.* **352**, 36–42.

Martin, J., Mayhew, M., Langer, T. & Hartl, F.-U. 1993 The reaction cycle of GroEL and GroES in chaperonin-assisted protein folding. *Nature, Lond.* **366**, 228–233.

McMullin, T.W. & Hallberg, R.L. 1988 A highly evolutionary conserved mitochondrial protein is selectively synthesized and accumulated during heat shock in Tetrahymena thermophila. *Molec. Cell Biol.* **7**, 4414–4423.

Ostermann, J., Horwich, A.L., Neupert, W. & Hartl, F.U. 1989 Protein folding in mitochondria requires complex formation with hsp60 and ATP hydrolysis. *Nature, Lond.* **341**, 125–130.

Palleros, D.R., Reid, K.L., Shi, L., Welch, W.J. & Fink, A.L. 1993 ATP-induced protein–HSP70 complex dissociation requires K⁺ but not ATP hydrolysis. *Nature, Lond.* **365**, 664–666.

Pelham, H.R.B. 1986 Speculations on the functions of the major heat shock and glucose-regulated proteins. *Cell* **46**, 959–961.

Robinson, C.V., Groβ, M., Eyles, S.J., Ewbank, J.E., Mayhew, M., Hartl, F.U., Dobson, C.M. & Radford, S. 1994 Hydrogen exchange protection in GroEL-bound α-lactalbumin detected by mass spectrometry. *Nature, Lond.* **372**, 646–651.

Saibil, H.R., Zheng, D., Roseman, A.M., Hunter, A.S., Watson, G.M.F., Chen, S., auf der Mauer, A., O'Hara, B.P., Wood, S.P., Mann, N.H., Barnett, L.K. & Ellis, R.J. 1993 ATP induces large quarternary rearrangements in a cage-like chaperonin structure. *Curr. Biol.* **3**, 265–273.

Schmidt, M., Rutkat, K., Rachel, R., Pfeifer, G., Jaenicke, R., Viitanen, P., Lorimer, G. & Buchner, J. 1994 Symmetric complexes of GroE chaperonins as part of the functional cycle. *Science, Wash.* **265**, 656–659.

Schröder, H., Langer, T., Hartl, F.U. & Bukau, B. 1993 DnaK, DnaJ and GrpE form a cellular chaperone machinery capable of repairing heat-induced protein damage. *EMBO J.* **12**, 4137–4144.

Staniforth, R.A., Burston, S.G., Atkinson, T. & Clarke, A.R. 1994 Affinity of chaperonin-60 for a protein substrate and its modulation by nucleotides and chaperonin-10. *Biochem. J.* **300**, 651–658.

Stuart, R.S., Cyr, D.M., Craig, E.A. & Neupert, W. 1994 Mitochondrial molecular chaperones: their role in protein translocation. *Trends Biochem. Sci.* **19**, 87–92.

Szabo, A., Langer, T., Schröder, H., Flanagan, J., Bukau, B. & Hartl, F.U. 1994 The ATP hydrolysis-dependent reaction cycle of the *Escherichia coli* Hsp70 system – DnaK, DnaJ and GrpE. *Proc. natn. Acad. Sci. U.S.A.* **91**, 10345–10349

Todd, M.J., Viitanen, P.V. & Lorimer, G.H. 1993 Hydrolysis of adenosine 5'-triphosphate by Escherichia coli GroEL: effects of GroES and potassium ion. *Biochemistry* **32**, 8560–8567.

Weissman, J.S., Kashi, Y., Fenton, W.A. & Horwich, A.L. 1994 GroEL-mediated protein folding proceeds by multiple rounds of binding and release of nonnative forms. *Cell* **78**, 693–702.

Zheng, X., Rosenberg, L.E., Kalousek, F. & Fenton, W.A. 1993 GroEL, GroES and ATP-dependent folding and sontaneous assembly of ornithine transcarbamylase. *J. biol. Chem.* **268**, 7489–7493.

Unliganded GroEL at 2.8 Å: structure and functional implications

PAUL B. SIGLER[1] AND ARTHUR L. HORWICH[2]

Howard Hughes Medical Institute and the Departments of [1] Molecular Biophysics and Biochemistry and of [2] Genetics, Yale University, 295 Congress Avenue, New Haven, Connecticut 06510, U.S.A.

SUMMARY

The three-dimensional structure of the *E. coli* chaperonin, GroEL, has been determined crystallographically and refined to 2.7 Å in two crystal forms: an orthorhombic form from high salt and a monoclinic form from polyethylene glycol. The former is ligand free, the latter is both liganded with ATP analogues and ligand free. These structures provide a structural scaffold upon which to interpret extensive mutagenesis and biochemical studies.

GroEL contains two sevenfold rotationally symmetric rings of identical 547-amino acid subunits. The rings are arranged 'back-to-back' with exact dyad symmetry to form a stubby cylinder that is 146 Å high with an outer diameter of about 143 Å. The cylinder has a substantial central channel that is unobstructed for the entire length of the cylinder and has a diameter of about 45 Å except for large bulges that lead into a sevenfold symmetric array of elliptical side windows in each ring.

Each subunit is composed of three distinct domains: (i) an 'equatorial' domain that contains the N- and C-terminus and the ATP-binding pocket, (ii) an 'apical domain' that forms the opening of the central channel and contains poorly ordered segments that mutational studies implicate in binding unfolded polypeptides and GroES, and (iii) an intermediate domain tht connects the other two domains and may serve to transmit allosteric adjustments.

1. INTRODUCTION

Our research work addresses the molecular mechanism of chaperonin-assisted protein folding by GroEL. Recent work, some of it presented in this volume, describe various activity cycles that include: (i) the cycling of ATP/ADP and the heptameric co-chaperonin, GroES (Ellis *et al.* 1991; Gething *et al.* 1992; Hendrick *et al.* 1993; Horwich *et al.* 1993; Todd *et al.* 1994); this cycle is necessary for GroEL function but can occur independently in the absence of any 'substrate' polypeptide; and (ii) various scenarios by which GroEL binds and releases polypeptide and, thereby, prevents (or possibly reverses) the formation of essentially irreversible off pathway folding intermediates (Martin *et al.* 1994; Todd *et al.* 1994; Weissman *et al.* 1994). Overall, GroEL, in concert with GroES and ATP, increases the efficiency by which linearly encoded information is converted into functional protein. We discuss here the three-dimensional structure of unliganded GroEL determined by X-ray crystallography to 2.8 Å, a resolution that enabled us to define most of the stereochemistry in near atomic detail (Braig *et al.* 1994). This structure, and a parallel study of directed mutational changes (Fenton *et al.* 1994), as well as on-going work on various liganded states, begins a structure–function analysis that should enable us to: (i) define the binding of substrates and cofactors in chemical terms; (ii) describe the molecular

mechanisms underlying the dramatic conformational differences observed in various liganded states; and (iii) provide a rational basis for the design and interpretation of future experiments examining dynamics of the reaction cycle. In short, we wish to create a sound framework for a truly biochemical understanding of chaperonin function.

2. THE OVERALL ARCHITECTURE

Figure 1 shows GroEL to be a nearly sevenfold rotationally symmetrical cylinder composed of two rings of seven subunits each. It is a bit taller (146 Å) than it is wide (137 Å). The rings are stacked 'back-to-back', related by dyad symmetry. There is a central channel in the cylinder that is about 45 Å wide both at its openings and at the level of the equatorial plane, that passes through the interface between the rings. This channel widens considerably at the mid-latitude of each ring, through bulges into each subunit that connect the central channel to the outside through seven 'side windows'.

The subunits form three domains (see figures 1 and 2). First, a well-ordered, highly α-helical, 'equatorial domain' of 243 residues (6–133, 409–523) that forms a solid well-organized foundation around the waist of the assembly. It provides most of the side-to-side contacts between subunits within the ring and all of the contacts

Phil. Trans. R. Soc. Lond. B (1995) **348**, 113–119
Printed in Great Britain

113

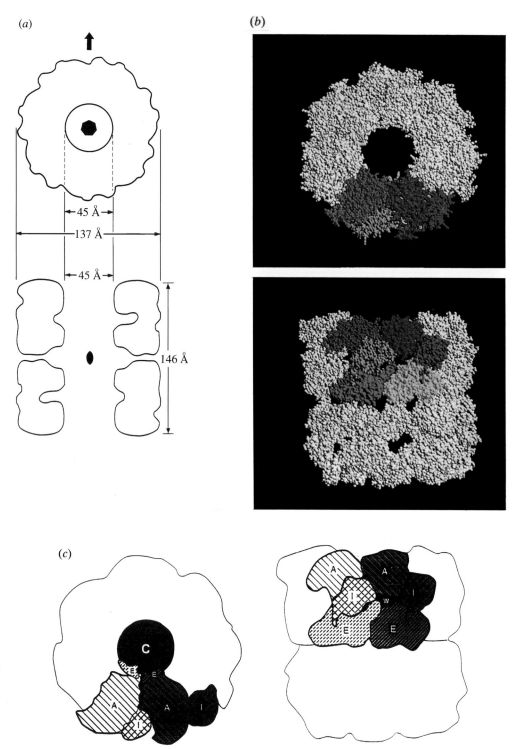

Figure 1. Overall structure of GroEL. (*a*) Dimensions of the tetradecamer. Top view looking down into the opening of the cylindrical channel and a sagittal cut in the plane of the cylindrical axis. Dyads indicated by (↑ and ●) and the sevenfold by (●). (*b*) Spacefilling model highlighting two adjacent subunits of the top ring. The domains are colour coded in the left subunit (green for equatorial; gold for intermediate and purple for apical) and in the right subunit, yellow for equatorial, red for intermediate, blue for apical. (*c*) Diagram of subunit and domain structure highlighting the same two subunits. E, I, A stand for equatorial, intermediate and apical. W is the external window of the channel that connects to the central channel. Reproduced with permission from Braig *et al.* (1994).

across the equatorial plane that hold the two rings together. Second, an 'apical domain' (191–376) of 186 residues that surrounds the openings at the top and bottom of the central channel. The apical domain's internal structure is less well defined and it is more loosely positioned and oriented in the assembly. Finally, a small slender 'intermediate domain' at the

periphery of the cylinder links the equatorial domain to the apical domain through well-ordered secondary structure elements. The connection between the intermediate and apical domains is a pair of antiparallel β strands that appears to form a hinge permitting global movements of the apical domain relative to the rest of the assembly. The connection to the equatorial domain is a more robust one, where α-helical elements project from the equatorial domain at a point that flanks the ATP binding site. The solvent compartment within the cylinder is a convoluted chamber consisting of the wide central channel with fourteen (seven in each ring) much narrower and irregular extensions reaching outward, underneath the apical domains, towards the intermediate domains and the side windows. Thus, even if the central channel is occupied, small solution components can enter or exit the chamber freely via the side windows. The surface of the solvent chamber is extensive and its contours would be altered significantly by changes in the orientation of the apical domain.

3. THE STRUCTURE DETERMINATIONS

Although the details of the specimen preparation, crystallization, structure determination and refinement have been published, we point out important short-comings that must be appreciated because they bear on the functional interpretation. In brief, the structure is based on a single good isomorphous derivative (SIR), and phase improvement derived from the sevenfold non-crystallographic symmetry (NCS). The initial experimental electron density was interpretable in terms of a nearly contiguous polypeptide chain with most of the distinctive side chains in register with the amino acid sequence and compatible with the mercury-atom sites attached to the cysteine residues. However, certain regions in the apical domain were not inter-pretable in terms of the sequence and some were not even contiguous. The original model was used only to build an envelope around each subunit, a procedure required by the averaging algorithm RAVE (Jones *et al.* 1991) which does not necessarily rely on a single sevenfold rotational symmetry operator. This was done by centering a sphere of 5 Å radius at each α-carbon. The original model was then discarded as were the SIR phases. Starting with 8 Å random phases and only the molecular envelope described above, new phases were created with the electron-density averaging program RAVE that allowed the NCS to depart from the original exact sevenfold rotational axis and to relate each of the subunits to a reference subunit through individual operators. Iterative phase extension and periodic upgrading of the NCS operators led to a map that was essentially the same as the original experimental map, thereby, ensuring its validity. There were, however, noticeable improvements in contiguity and side-chain identity in the apical domain which suggested small but significant departures from sevenfold symmetry especially in the apical domain. Indeed, significant discontinuities and ambiguities in side chain assign-ment – all in the apical domain – were not resolved by the refinement (see figure 2 of Braig *et al.* 1994). Moreover, the violations of backbone torsional restraints (φ, ϕ) and a sudden dip in the three-dimensional profile (Luthy *et al.* 1992) of the segment between 295–315 suggests a mistaken interpretation of a loop extending from across the top of the apical domain towards the central cavity.

Because the mutations that disrupt polypeptide binding are all in the apical domain, and some of them, such as those at 234, 237, 259, 263, 264 are in these poorly defined loops (figures 2 and 3), it is tempting to speculate that the poorly defined surface segments of the apical domain are designed to be flexible enough to bind a wide range of polypeptide sequences. This explanation is supported by the following observations:

(1) The large temperature factors that characterize most portions of the apical domain give way to well-ordered structure in the apical segments that contact the neighbouring intermediate domain. This implies a local rather than a global disordering of the apical domain.

(2) The low temperature factors and the high quality of the (2 Fo-Fc) maps in the intermediate and equatorial domains suggest that the inability to model certain apical segments is not due to inaccurate phases.

(3) A preliminary refined model of ATP-bound GroEL in a different monoclinic, crystal form shows essentially the same trace of the polypeptide and a tendency for the same apical segments to be poorly represented in the electron density (D. Boisvert *et al.*, unpublished).

On the other hand, we cannot confidently ascribe local flexibility to these poorly defined segments until we significantly improve the quality and quantity of high resolution data. In addition to the apical segments discussed above, the N-terminal 5 residues and C-terminal 26 residues are not visualized. The poly-peptide segments contiguous to these terminal sequence segments are well defined components of the equatorial domain and suggest that these terminal peptides project into the central cavity in a crystallographically disordered manner. Reconstructions of electron micro-graphs (Saibil *et al.* 1993; Chen *et al.* 1994) show density in the central cavity at the level of the equatorial domains which may reflect the accumu-lation of 217 residues from the N and C terminal segments of seven subunits.

4. FUNCTIONAL ISSUES
(a) *Relationships between subunits*

The subunits within the ring are held together primarily by contacts between the equatorial domains. Figure 2 shows that one of these contacts is formed by a stem-loop of β-structure that reaches out against the channel surface of the neighbouring equatorial domain forming parallel β-structure with segment 519–522 that just precedes the disordered C-terminus. Mutational analysis is consistent with this picture since C-terminal deletions are tolerated, but only as far as 522; those that include 521 or additional residues are lethal because of GroEL disassembly (Burnett *et al.* 1994). Recently, mutations changing residues that form the contact surface between rings (see figure 4 of Braig *et al.* 1994) were shown to decouple the cylindrical assembly into separate stable seven subunit rings (J. Weissman *et al.* unpublished results). In the wild-type molecule, the seven equatorial domains of each ring face one another to form, in total, a large solvent excluded contact surface (5850 Å²). In contrast

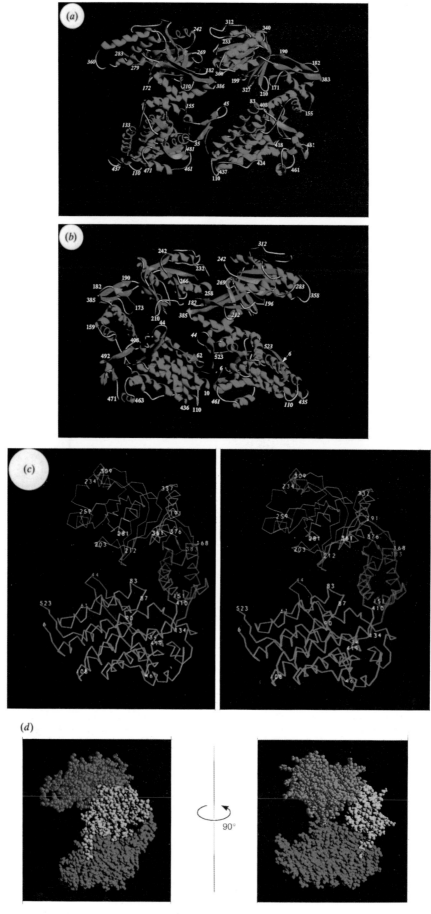

Figure 2. For description see opposite.

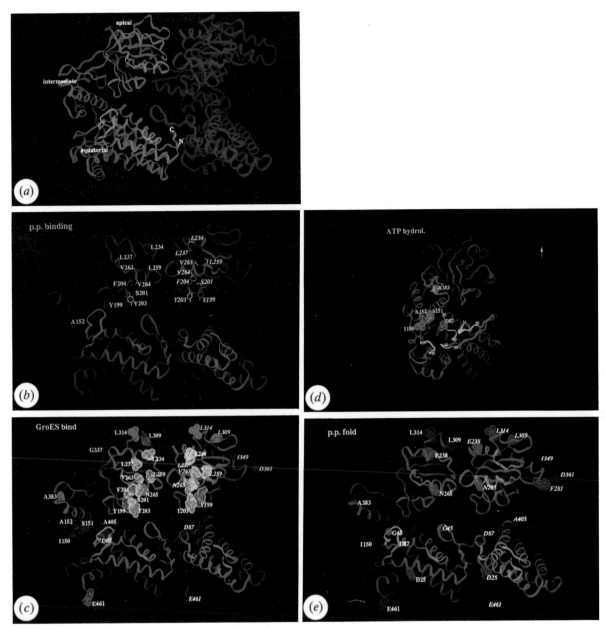

Figure 3. Amino acids involved in GroEL functions. Shown in each panel are two subunits from the 'top' ring of GroEL, viewed from inside (i.e. from central channel). The ATP hydrolysis panel is an exception: the dimer has been rotated horizontally to view the binding pocket from the 'right' side. In panel (*c*) residues whose mutational alteration affected both polypeptide and GroES binding are shown in yellow spheres; other residues whose alteration affected GroES binding but not polypeptide binding are shown in blue spheres. In panel (*d*) residues affecting ATP hydrolysis are shown in green spheres. Conserved residues around the binding pocket are shown by yellow ribbon. (*e*) Shows residues whose alteration interfered with release or folding of polypeptide. Ribbon diagram of main chain derived from the averaged model (Braig *et al.* 1994), displayed in InsightII (BioSym). Reproduced from Braig *et al.* (1994) with permission.

to the contacts within the rings, the complementary surfaces do not interdigitate across the equatorial plane.

The intermediate domain emerges from the equatorial domain adjacent to the ATP-binding pocket and extends diagonally up and to the right (when the top ring is viewed from the outside as in figures 1 and 2), where it contacts the apical domain of the neighbouring subunit. As the intermediate domain contacts the neighbouring apical domain and, of course, is

Figure 2. (*a*) Ribbon drawing of two adjacent subunits in the top ring (see figure 1) viewed from the outside. Beta-strands are wide green arrows, alpha-helices are blue, and Connecting Strands are yellow tubes. (*b*) The same two subunits viewed from the inside of the channel. (*c*) Stereopair of the alpha carbon backbone of a subunit oriented as in the right panel of (*d*). The colour coding is green, blue and red for the equatorial, intermediate and apical domains, respectively. (*d*) Two spacefilling models of the left highlighted subunit in figure 1 as viewed from the outside (left panel) and then rotated 90° so that the outside surface faces towards the right (right panel). Reproduced with permission from Braig *et al.* (1994).

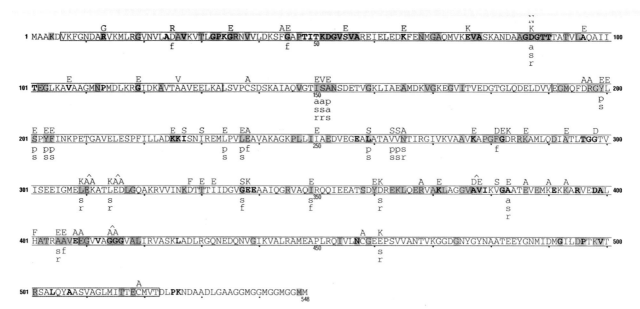

Figure 4. Amino acid substitutions in GroEL and the functions affected shown above and below, respectively, the wild-type amino-acid sequence; a are defective in ATP binding, f are defective in folding, p are defective in peptide binding, r are defective in release and s are defective in GroES binding. Residues conserved in all chaperonins (Hsp60/GroEL family and TF55/TCP1 family) are shown in bold. Residues in at least 49 or 50 Hsp60/GroEL chaperonin shown in orange background.

linked covalently to the apical domain of the same subunit, it provides a structural link that can allosterically couple stereochemical changes in the equatorial domain associated with nucleotide binding and hydrolysis to tertiary and quaternary changes of the apical region. The view that the intermediate domain acts as an allosteric lever involved in the coupling of quaternary changes to the enzymatic cycling of ATP and ADP is supported by a mutational change that breaks an ionic interaction between the apical domains (Arg 197) and neighbouring intermediate domains (Glu 386) and, thereby, disrupts positive cooperativity in ATPase action within the ring and the negative cooperativity between the rings (Yifrach *et al.* 1994). Although the structure and genetics provide strong clues to allosteric effects within the ring, it is less clear how enzymic and binding events influence structure across the equatorial plane in the opposite ring.

(b) Peptide binding

Fenton *et al.* (1993) have reported mutations that disrupt binding of polypeptide substrates (figure 3). Although an exhaustive analysis has not been completed, it appears that all of the changes that inhibit the binding of non-native polypeptides (without disruption of the GroEL assemblies) affect residues that are hydrophobic in character mapping within the apical domain. All of these mutational changes are at the surface and either face towards the central channel or lie on the domains' undersurface which forms the 'roof' of the outward evagination of the central channel. Such residues may form hydrophobic contacts with exposed hydrophobic surfaces of nonnative intermediates. Most interesting is the fact that all of these mutations also disrupt binding of GroES (figure 3). In sum, from a structural vantage point, the hinge-like

attachment of the apical domain, the allosterically responsive contact to the neighbouring intermediate domain and the porosity of the assembly that allows room for structural adjustment are structural features seemingly designed to potentiate significant structural change in the apical domain and, thereby, present a range of potentially flexible binding surfaces to polypeptides and GroES,

Despite such flexibility, the crystal structure provides dimensional constraints on GroEL's capacity to bind non-native polypeptide chains. From electron microscopic (Braig *et al.* 1993; Langer *et al.* 1993; Saibil *et al.* 1993, 1994; Chen *et al.* 1994) and mutation studies (Fenton *et al.* 1994), it is likely that a significant portion of the bound polypeptide is confined to the central cavity. The 45 Å opening at the top and bottom of the channel defines a cylindrical space of $\sim 250\,000$ Å^3 from one opening to the other. A native protein in a crystal lattice occupies about 2.4 Å^3 per Dalton which means the entire channel could accept an elongated ellipsoidal protein of ~ 100 kDa provided it could cross the equatorial plane. GroEL does not bind fully folded native proteins but is thought to bind non-native intermediates with native-like secondary structure and variable amount of tertiary structure (Hayer–Hartl *et al.* 1994; Okazaki *et al.* 1994; Robinson *et al.* 1994; Zahn *et al.* 1994). A classical 'molten globule' conformation (Okazaki *et al.* 1994), while it may turn out to be more collapsed than what is generally recognized, already occupies 20 to 30% more volume than the native structure, meaning ~ 75–80 kDa could be accommodated in the entire 45 Å diameter cylindrical channel. As only one side of the double toroid is thought to bind polypeptide at one time (Hendrick *et al.* 1993; Horwich & Willison 1993; Todd *et al.* 1994), this figure would be halved to 38–40 kDa per ring.

It is particularly surprising that residues, such as Tyr 199 and Tyr 203 (figure 3), located deep within the central channel on a re-entrant surface of the apical domain, are mutationally implicated in both peptide binding and GroES binding. Whereas, peptides almost certainly bind in the central channel and may reach these deeply situated residues by either a further unfolding or adjustments in orientation of the apical domain, the case for the interaction of native GroES with such residues is not so clear. GroES could fit into the central channel of GroEL. Recent cryoelectromicrograph studies of Saibil and coworkers, however, of the GroEL–GroES binary complex, suggest that the GroEL apical domain undergoes a conformational change that extends it upward to meet the edges of a dome-shaped GroES 7-mer (Saibil *et al.* 1993; Chen *et al.* 1994). The central face of the apical domains including the residues implicated in peptide binding thus appear to be able to become much more accessible to contact with GroES. This raises the interesting question of whether GroES and peptide compete for the same binding sites on the same side of GroEL, or whether the two ligands generally interact at opposite ends of the cylinder. This is a major question that must be resolved experimentally.

5. GroES BINDING, ATP BINDING AND HYDROLYSIS, AND POLYPEPTIDE RELEASE

Figure 4 summarizes the residues where mutational changes disrupt key functions. For many residues, there is a noticeable overlap (Fenton *et al.* 1994). For example, GroES binding requires all of the same hydrophobic residues on the inside surface of the apical domain that are required for polypeptide binding. However, GroES also requires two non-polar residues on the outer top surface of the apical domain, as well as, other polar and non-polar residues throughout the intermediate and equatorial domains (one of which forms a contact between rings!). Many of these same residues are also implicated in ATP binding and hydrolysis as well as peptide release. ATP hydrolysis is sensitive to changes in the intermediate domain and to one change (involving D87) in the ATP-binding pocket of the equatorial domain. The location of such mutations reinforces the view that the intermediate domain is involved in the allosteric modulations of the ATPase function.

Finally, it should be stressed that the overlapping distribution of mutations affecting the binding and release of nucleotide, polypeptide and GroES as well as the hydrolysis of ATP underscores the interdependence of such functions. For a more precise understanding of the chemistry of this interdependence, we must visualize the various liganded states of GroEL. This work is in progress.

We acknowledge the contributions to this work of our colleagues: Dr Z. Otwinowski, Dr K. Braig, Dr R. Hegde, Dr D. Boisvert, Dr A. Joachimiak, Dr W. Fenton, Dr Y. Kashi and Dr K. Furtak. In addition, we thank Dr J. Weissmann, Dr G. Lorimer and Dr H. Saibil for stimulating discussion.

REFERENCES

Braig, K.B., Otwinowski, Z., Hegde, R., Boisvert, D.C., Joachimiak, A., Horwich, A.L. & Sigler, P.B. 1994 *Nature, Lond.* **371**, 578–586.

Braig, K., Simon, M., Furuya, F., Hainfeld, J.F., & Horwich, A.L. 1993 *Proc. natn. Acad. Sci. U.S.A.* **90**, 3978–3982.

Burnett, B., Horwich, A.L. & Low, K.B. 1994 *J. Bact.* **176**, 6980–6985.

Chen, S., Roseman, A.M., Hunter, A.S., Wood, S.P., Burston, S.G., Ranson, N.A., Clarke, A.R. & Saibil, H.R. 1994 *Nature, Lond.* **371**, 261–264.

Ellis, R.J. & van der Vies, S.M. 1991 *A. Rev. Biochem.* **60**, 321–347.

Fenton, W.A., Kashi, Y., Furtak, K. & Horwich, A.L. 1993 *Nature, Lond.* **371**, 614–619.

Gething, M.J. & Sambrook, J. 1992 *Nature, Lond.* **355**, 33–45.

Hayer-Hartl, M.K., Ewbank, J.J., Creighton, T.E. & Hartl, F.U. 1994 *EMBO J.* **13**, 3192–3202.

Hendrick, J.P. & Hartl, F.-U. 1993 *A. Rev. Biochem.* **62**, 349–384.

Horwich, A.L. & Willison, K. 1993 *Phil. Trans. R. Soc. Lond.* B **339**, 313–326.

Jones, T.A. *et al.* 1991 *Acta Crystallogr.* **A47**, 110–119.

Langer, T., Pfeifer, G., Martin, J., Baumeister, W. & Hartl, F.U. 1992 *EMBO J.* **13**, 4757–4765.

Luthy, R., Bowie, J.U. & Eisenberg, D. 1992 *Nature, Lond.* **356**, 83–85.

Martin, J., Mayhew, M., Langer, T. & Hartl, F.U. 1993 *Nature* **366**, 228–233.

Okazaki, A., Ikura., T., Nakaido, K. & Kuwajima, K. 1994 *Struct. Biol.* **1**(7), 439–445.

Robinson, C.V., Broβ, M., Eyles, S.J., Ewbank, J.J., Mayhew, M., Hartl, F.U., Dodson, C. & Radford, S.E. 1994 *Nature, Lond.* **372**, 646–651.

Saibil, H.R. *et al.* 1993 *Curr. Biol.* **3**, 265–273.

Saibil, H.R. 1994 *Nature Struct. Biol.* **1**, 838–842.

Todd, M.J., Viitanen, P.V. & Lorimer, G.H. 1994 *Science, Wash.* **265**, 659–666.

Weissman, J.S., Kashi, Y., Fenton, W.A. & Horwich, A.L. 1994 *Cell* **78**, 693–702.

Yifrach, O. & Horovitz, A. 1994 *J. molec. Biol.* **243**, 397–401.

Zahn, R., Spitzfaden, C., Ottiger, M., Wuthrich, K. & Pluckthun, A. 1994 *Nature, Lond.* **368**, 261–265.